AF886255

SOTHIS PRESS

Sothis Press aims to completely reimagine the public domain book with an emphasis on universal access to captivatingly designed print editions. We believe that technology will enable every single book ever published to be perpetually accessible in beautiful print form. Sothis is beauty in ubiquity.

www.sothis.press
1 Waterhouse Square, London, ECIN 2ST

Rights of Trains on Single Track

A Complete Examination for Employes
on the Standard Code and Other
Recommended Train Rules

Harry Willard Forman

SOTHIS PRESS

ISBN 978-1-0365-0191-4

This edition is published by
Sothis Press © 2023. The company
is registered in England and Wales.
The Company Number is 14900389.

RIGHTS OF TRAINS

By the late

HARRY W. FORMAN

Assistant to Vice-President and General Manager, Western Pacific Railroad

Revised by

PETER JOSSERAND

Night Chief Dispatcher, Western Pacific Railroad, Sacramento Calif.; Rules Department Editor, Brotherhood of Locomotive Engineers Journal; Rules Consultant, Brotherhood of Locomotive Firemen and Enginemen's Magazine

FOURTH EDITION

SIMMONS-BOARDMAN PUBLISHING CORP.

New York, N. Y.

FOREWORD TO FOURTH EDITION

More than one hundred years have elapsed since American railroads first began operation of more than one engine on a given stretch of track. With the coming of the second engine, provisions had to be made to prevent opposing movements from colliding, and it was at this early time that the first operating rules came into existence. These rules were at first comparatively simple, but, as traffic increased, they became progressively more complicated.

There was no coordinated effort to produce uniform rules on the various roads at first. Each road contrived whatever rules it deemed necessary.

In the late 1880's, the industry became conscious of the need for pooling their hard-won knowledge and experience to produce the best possible set of operating rules, to be made available to all roads. Much of this knowledge had arisen out of hazard which had resulted from inadequate rules.

In 1889 the General Time Convention adopted the first STANDARD CODE, designated as *Uniform Train Rules and Rules for the Movement of Trains by Telegraphic Orders.* It represented the accumulated knowledge of the best operating department minds in the industry. Out of the General Time Convention developed the Association of American Railroads.

This first effort was not enough. Some of the rules were not "air-tight." There developed hazards which necessitated further revision. Conditions on the railroads changed; the density of traffic increased; telephones were brought into use for dispatching; multiple tracks replaced single-track operation in places; and there were many other changes too numerous to mention. To keep pace with progress in the industry, as well as to plug loopholes in the rules where necessary, many revisions of the Standard Code have been made during the past sixty years.

The Standard Code today is the foundation of all rule books and the yardstick by which safety is measured on American railroads. Since operating conditions on the individual carrier may make it desirable to vary from the Standard Code to produce the most flexible performance, the Code provides that where proper safeguards are established, this may be done. This makes for a difference in rules from road to road and identical rule books are seldom found. For this reason, no treatise such as this can be cited as an absolute authority on the rules of all railroads.

In this Fourth Edition, it has been my aim to set forth clearly the basic principles underlying the rules, and to show how the rules of various roads differ from the Standard Code and from each other.

The beginner should obtain a broad perspective of the rules so as to develop ability to apply his knowledge to the rules of his individual road in solving such problems as may arise in his work. Once the basic principles are understood it is not difficult to apply them to any given set of rules.

There is another condition, existing to a lesser degree, of which the reader should be cognizant. Interpretations of an identical rule may vary from one road to another. Where a Rules Examiner, backed by the Rules Committee, decides to alter the interpretation of a rule from that generally accepted by other roads, said interpretation becomes the law unto that particular road. There is no court of appeals from such a decision. THE RULES EXAMINER IS ALWAYS RIGHT.

The individual company's book of rules is composed of terse statements, reduced to a minimum of words. It is doubtful if the most intelligent nonrailroader could, unassisted, study the average book of rules and successfully pass a rules examination. This book provides assistance to those studying train operating rules, by clarifying the meaning of the cryptic rules which railroads expect employes to memorize, or learn by rote.

The late Harry W. Forman, assistant to the vice-president and general manager of the Western Pacific Railroad, was one of the first to recognize the need for a book of this kind. His First Edition, *Rights of Trains on Single Track,* was privately published in 1904. Recognized as a national authority on train rules, he developed the principles of many of our present-day rules. It is with humbleness of spirit that I have stepped into the shoes of this great rules man to carry on his work.

Because of an ever-increasing percentage of rail lines being operated under some form of signal control, this edition has been expanded to include Block Signal and Interlocking Rules, in addition to Operating Rules. These rules will be found to differ from the Standard Code in some cases and to vary from road to road. Also, types of signals may vary from the semaphore of ancient vintage to the Pennsylvania's color-position light signals.

The rapid adoption of Centralized Traffic Control has necessitated the writing of special rules which are not uniform. "Traffic control," as it has been officially designated to get away from the trade name, is tailor-made to fit the exact requirements of each property and the rules are altered to fit the installation. The Association of American Railroad Superintendents, after considerable study, has recommended a set of traffic control rules to the Rules Committee of the Association of American Railroads for inclusion in the Standard Code.

Safety in railroad operation is directly dependent on each employe being conversant with and obeying the rules. "Safety" includes the lives of employes, as well as passengers and railroad property.

Any railroad, with extremely rare exceptions, can be operated free from accidents attributable to "failure of the human element" if its employes know and obey the rules. This holds true even when block signals and other safety devices are absent.

Conversely, no railroad can be operated safety if its employes do not know and obey the rules, no matter how elaborately it may be equipped with safety devices. These devices, per se, do not produce the safety factor. Rules give the devices meaning. Employes must know and obey the rules before the safety devices are effective.

There is no industry where little mistakes can have such far-reaching, and often disastrous, results. Safety is of the first importance in the discharge of duty; and safety arises out of a thorough knowledge of and obedience to the rules.

In conclusion, I wish to express my thanks to the officers of the Association of American Railroads for their assistance; and to the many operating officers of those railroads listed below, whose contributions in time and material over a period of years have made this work possible. I am especially indebted to C. I. Morton, superintendent, Seaboard Air Lines, and president of the American Railroad Superintendents Association, who read the manuscript for possible omissions and errors: to W. R. Triem, general superintendent of telegraph, Pennsylvania Railroad, and chairman of the Rules Committee of the Association of American Railroads; and to my good friend, Grant S. Allen, of the Western Pacific.

Those roads whose rules have been made available for a comparative study are:

Foreword

Atchison Topeka and Santa Fe
Baltimore and Ohio
Chicago Burlington and Quincy
Chicago Milwaukee and St. Paul
Chicago Northwestern
Canadian Pacific
Great Northern
Illinois Central
Missouri Pacific
Northern Pacific
New York Central
Texas and Pacific
Pennsylvania
Seaboard Air Line
Southern Pacific
Western Pacific

PETER JOSSERAND
Sacramento, California
September, 1951

CONTENTS

RIGHTS OF TRAINS

Part I

Chapter I

THE STANDARD CODE OF OPERATING RULES, BLOCK SIGNAL RULES AND INTERLOCKING RULES*

Adopted March, 1949

WHERE PROPER SAFEGUARDS ARE PROVIDED, RAILROADS MAY MODIFY OR ADD TO THESE RULES TO SUIT THEIR REQUIREMENTS

General Note.—Blanks in the rules may be filled by each railroad to suit its own organization or requirements.

(Form of order putting rules in effect.)

The rules herein set forth govern the railroads operated by the .... Company. They take effect ...., superseding all previous rules and instructions inconsistent therewith.

Special instructions may be issued by proper authority.

(Name) ....

(Title) ....

GENERAL NOTICE

Safety is of the first importance in the discharge of duty.

Obedience to the rules is essential to safety.

To enter or remain in the service is an assurance of willingness to obey the rules.

---

*Reproduction in full courtesy of the Association of American Railroads, Operations and Maintenance Department, Operating—Transportation Division, Operating Section.

The service demands the faithful, intelligent and courteous discharge of duty.

To obtain promotion, ability must be shown for greater responsibility.

GENERAL RULES

A. Employes whose duties are prescribed by these rules must provide themselves with a copy.

Employes whose duties are in any way affected by the time-table must have a copy of the current time-table with them while on duty.

B. Employes must be conversant with and obey the rules and special instructions. If in doubt as to their meaning, they must apply to proper authority for an explanation.

C. Employes must pass the required examinations.

D. Persons employed in any service on trains are subject to the rules and special instructions.

E. Employes must render every assistance in their power in carrying out the rules and special instructions and must report to the proper official any violation thereof.

F. Accidents, failure in the supply of water or fuel, defects in the track, bridges, signals, or any unusual conditions which may effect the movement of trains, must be promptly reported by quickest available means to the proper authority.

G. The use of intoxicants or narcotics by employes subject to duty, or their possession or use while on duty, is prohibited.

H. The use of tobacco by employes on duty in or about passenger stations, or on passenger cars, is prohibited.

J. Employes on duty must wear the prescribed badge and uniform and be neat in appearance.

K. To avoid annoyance to the public, employes and others authorized to transact business at stations, and on or about trains, must be courteous, orderly and quiet.

L. In case of danger to the Company's property, employes must unite to protect it.

M. Employes must exercise care to avoid injury to themselves or others. They must observe the condition of equipment and the tools which they use in performing their duties, and when found defective will, if practicable, put them in safe condition, reporting defects to the proper authority.

They must inform themselves as to the location of structures or obstructions where clearances are close.

They must expect the movement of trains, engines or cars at any time, on any track, in either direction.

They must not stand on the track in front of an approaching engine or car for the purpose of boarding the same.

DEFINITIONS

OPERATING RULES—BLOCK SIGNAL RULES—INTERLOCKING RULES

APPROACH SIGNAL—A fixed signal used in connection with one or more signals to govern the approach thereto.

AUTOMATIC BLOCK SYSTEM—A series of consecutive blocks governed by block signals, cab signals, or both, actuated by a train, or engine, or by certain conditions affecting the use of a block.

BLOCK—A length of track of defined limits, the use of which by trains and engines is governed by block signals, cab signals, or both.

BLOCK SIGNAL—A fixed signal at the entrance of a block to govern trains and engines entering and using that block.

BLOCK STATION—A place from which block signals are operated.

BLOCK SYSTEM—A series of consecutive blocks.

CAB SIGNAL—A signal located in engineman's compartment or cab, indicating a condition affecting the movement of a train or engine and used in conjunction with interlocking signals and in conjunction with or in lieu of block signals.

CURRENT OF TRAFFIC—The movement of trains on a main track, in one direction, specified by the rules.

DIVISION—That portion of a railroad assigned to the supervision of a ...., (usually superintendent).

DWARF SIGNAL—A low home signal.

ENGINE—A unit propelled by any form of energy, or a combination of such units, operated from a single control, used in train or yard service.

EXTRA TRAIN—A train not authorized by a timetable schedule. It may be designated as—

Extra—for any extra train except passenger extra or work extra;
Passenger Extra—for passenger train extra;
Work Extra—for work train extra.

FIXED SIGNAL—A signal of fixed location indicating a condition affecting the movement of a train or engine.

HOME SIGNAL—A fixed signal at the entrance of a route or block to govern trains or engines entering and using that route or block.

INTERLOCKING—An arrangement of signals and signal appliances so interconnected that their movements must succeed each other in proper sequence and for which interlocking rules are in effect. It may be operated manually or automatically.

INTERLOCKING LIMITS—The tracks between the home signals of an interlocking.

INTERLOCKING SIGNALS—The fixed signals of an interlocking.

INTERLOCKING STATION—A place from which an interlocking is operated.

LIMITED SPEED—A speed not exceeding . . . . miles per hour.

MAIN TRACK—A track extending through yards and between stations, upon which trains are operated by time-table, or train order, or both, or the use of which is governed by block signals.

MANUAL BLOCK SYSTEM—A series of consecutive blocks, governed by block signals operated manually, upon information by telegraph, telephone or other means of communication.

MEDIUM SPEED—A speed not exceeding . . . . miles per hour.

PILOT—An employe assigned to a train when the engine-

NOTE TO DEFINITION OF FIXED SIGNAL.—The definition of a "Fixed Signal" covers such signals as switch, train order, block, interlocking, semaphore, disc, stop signs, yard limit signs, slow signs, or other means for displaying indications that govern the movement of a train or engine.

man or conductor, or both, are not fully acquainted with the physical characteristics or rules of the railroad, or portion of the railroad over which the train is to be moved.

REDUCED SPEED—Proceed prepared to stop short of train or obstruction.

REGISTER STATION—A station at which train register is located.

REGULAR TRAIN—A train authorized by a time-table schedule.

RESTRICTED SPEED—Proceed prepared to stop short of train, obstruction, or switch not properly lined and to look out for broken rail.

SCHEDULE—That part of a time-table which prescribes class, direction, number and movement for a regular train.

SECTION—One of two or more trains running on the same schedule displaying signals, or for which signals are displayed.

SIDING—A track auxiliary to the main track for meeting or passing trains.

SINGLE TRACK—A main track upon which trains are operated in both directions.

SLOW SPEED—A speed not exceeding .... miles per hour.

SPRING SWITCH—A switch equipped with a spring mechanism arranged to restore the switch points to normal position after having been trailed through.

STATION—A place designated on the time-table by name.

SUBDIVISION—A portion of a division designated by time-table.

SUPERIOR TRAIN—A train having precedence over another train.

TIME-TABLE—The authority for the movement of regular trains subject to the rules. It contains the classified schedules with special instructions relating to the movement of trains.

TRAIN—An engine or more than one engine coupled, with or without cars, displaying markers.

TRAIN REGISTER—A book or form which may be used at designated stations for registering signals displayed, the time of arrival and departure of trains and such other information as may be prescribed.

TRAIN OF SUPERIOR CLASS—A train given precedence by time-table.

TRAIN OF SUPERIOR DIRECTION—A train given precedence in the direction specified by time-table as between opposing trains of the same class.

TRAIN OF SUPERIOR RIGHT—A train given precedence by train order.

TWO OR MORE TRACKS—Two or more main tracks upon any of which the current of traffic may be in either specified direction.

YARD—A system of tracks within defined limits provided for the making up of trains, storing of cars and other purposes, over which movements not authorized by time-table, or by train order, may be made, subject to prescribed signals and rules, or special instructions.

YARD ENGINE—An engine assigned to yard service.

YARD SPEED—A speed that will permit stopping within one-half the range of vision.

CHAPTER II

OPERATING RULES

NOTE: Rules with a prefix "S" are for single track; those with a prefix "D" are for two or more tracks. Rules without a prefix are for single and two or more tracks. The prefixes "S" and "D" to be printed in italics.

ADOPTED MARCH, 1949

STANDARD TIME

1. Standard Time obtained from .... observatory will be transmitted to all points from designated offices at .... m. daily.

2. Each conductor, engineman and other employe as may be designated, must carry, while on duty, a reliable railroad grade watch, for which there must be a prescribed certificate on file with the railroad.

3. Unless otherwise provided, watches of conductors, enginemen and .... must be compared with a standard clock when commencing each day's work. The time when watches are compared must be registered on a prescribed form. The location of standard clocks will be shown in the time-table.

TIME-TABLES

4. Each time-table, from the moment it takes effect, supersedes the preceding time-table, and its schedules take effect on any division, or subdivision, at the

leaving time at their initial stations on such division, or subdivision. But when a schedule of the preceding time-table corresponds in number, class, day of leaving, direction, and initial and terminal stations with a schedule of the new time-table, a train authorized by the preceding time-table will retain its train orders and assume the schedule of the corresponding number of the new time-table.

Schedules on each division, or subdivision, date from their initial stations on such division, or subdivision.

Not more than one schedule of the same number and day shall be in effect on any division, or subdivision.

5. Not more than two times are given for a train at any station; where one is given, it is, unless otherwise indicated, the leaving time; where two, they are the arriving and the leaving times.

Unless otherwise specified, the time applies:

On single track, at the switch where an opposing train clears; where there is no switch it applies at the station.

On two or more tracks, at the station.

Schedule meeting or passing stations are indicated by figures in full-faced type, with the numbers of the trains to be met or passed in small figures adjoining.

6. The following letters when placed before the figures of the schedule indicate:

L—leave;

A—arrive;

s—regular stop;

f—flag stop to receive or discharge traffic.

SIGNALS AND THEIR USE

7. Employes whose duties may require them to give signals, must provide themselves with the proper appliances, keep them in good order and ready for immediate use.

8. Flags of the prescribed color must be used by day, and lights of the prescribed color and type by night.

9. Day signals must be displayed from sunrise to sunset, but when day signals cannot be plainly seen, night signals must be used in addition.
Night signals must be displayed from sunset to sunrise.

10. **COLOR SIGNALS**

COLOR	INDICATION
(a) Red.	Stop.
(b) ————	Proceed at reduced speed, and for other uses prescribed by the Rules.
(c) ————	Proceed, and for other uses prescribed by the Rules.
(d) White.	Flag stop (See Rule 28) and for other uses prescribed by the rules.
(e) Blue.	See Rule 26.
(f) Purple.	Stop. (Indication for siding derails.)

11. A fusee burning red on or near the track of an approaching train must be extinguished. The train may then proceed at reduced speed.

12. **HAND, FLAG AND LAMP SIGNALS**

NOTE: The hand, or a flag, moved the same as the lamp, as illustrated in the following diagrams, gives the same indication.

12 (a)

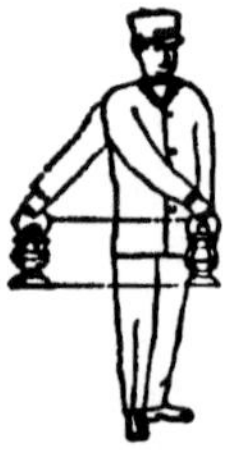

STOP
Swung at right angle to the track.

12 (b)

REDUCE SPEED
Held horizontally at arm's length.

12 (c)

PROCEED
Raised and lowered vertically.

12 (d)

BACK
Swung vertically in a circle at right angle to the track.

12 (f)

APPLY AIR BRAKES
Swung horizontally above the head when standing.

12 (g)

RELEASE AIR BRAKES
Held at arm's length above the head when standing.

12 (h) Any object waved violently by any one on or near the track is a signal to stop.

12 (i) Signals must be given from a point where they may be plainly seen and in such a way that they cannot be misunderstood. If there is doubt as to the meaning of a signal, or for whom it is intended, it must be regarded as a stop signal. If signals disappear from view, the movement must be stopped immediately.

14. ENGINE WHISTLE SIGNALS

NOTE: The signals prescribed are illustrated by "o" for short sounds: "–" for longer sounds. The sound of the whistle should be distinct, with intensity and duration proportionate to the distance signal is to be conveyed.

SOUND	INDICATION
(a) o	Apply brakes. Stop.
(b) — —	Release brakes. Proceed.
(c) — o o o	Flagman protect rear of train.
(d) — — — —	Flagman may return from west or south.§
(e) — — — — —	Flagman may return from east or north.§
(g) o o	Answer to 14 (k) or any signal not otherwise provided for.
(h) o o o	When standing, back. Answer to 12 (d) and 16 (c). When running, answer to 16 (d).
(j) o o o o	Call for signals.
(k) — o o	(SINGLE TRACK) To call attention of engine and train crews of trains of the same class, inferior trains and yard engines, and of trains at train order meeting or waiting points, to signals displayed for a following section. If not answered by a train, the train displaying signals must stop and ascertain the cause unless otherwise provided. (TWO OR MORE TRACKS) To call attention of engine and train crews and yard engines, to signals displayed for a following section.
(l) — — o —	Approaching public crossings at grade. To be prolonged or repeated until crossing is reached.
(m) ————	Approaching stations, junctions, railroad crossing at grade and ————, as may be required.
(n) — — o	Approaching meeting or waiting points. See Rule S-90.
(o) o —	Inspect train line for leak or for brakes sticking.

SOUND	INDICATION
(p) Succession of short sounds.	Alarm for persons or live stock on the track.
(q) — o	Where there are two main tracks on which movements are made in either direction by signal indications, trains on left track will sound this signal preceding signal 14 (d) or 14 (e). When running against the current of traffic: (1) Approaching stations, curves, or other points where view may be obscured. (2) Approaching passenger or freight trains and when passing freight trains. (3) Preceding the signals prescribed by (d), (e), (r), (s), (t), (u), (v) and (w).

For additional tracks the following signals may be used:

SOUND	INDICATION
(r) — — — — — o	Flagman may return from east or north on ———— track.§
(s) — — — — o	Flagman may return from west or south on ———— track.§
(t) — — — — — o o	Flagman may return from east or north on ———— track.§
(u) — — — — o o	Flagman may return from west or south on ———— track.§
(v) — — — — — o o o	Flagman may return from east or north on ———— track.§
(w) — — — — o o o	Flagman may return from west or south on ———— track.§

§—As prescribed by Rule 99.

15. The explosion of two torpedoes is a signal to proceed at reduced speed. The explosion of one torpedo will indicate the same as two, but the use of two is required.

Torpedoes must not be placed at stations or on public crossings.

16. COMMUNICATING SIGNALS

NOTE: The signals prescribed are illustrated by "o" for short sounds; "–" for longer sounds.

SOUND	INDICATION
(a) o o	When standing—start.
(b) o o	When running—stop at once.
(c) o o o	When standing—back.
(d) o o o	When running—stop at next passenger station.
(e) o o o o	When standing—apply or release air brakes.
(f) o o o o	When running—reduce speed.
(g) o o o o o	When standing—recall flagman.
(h) o o o o o	When running—increase speed.
(j) o o o o o o	When running—increase train heat.
(k) o — o	Shut off train heat.
(l) ————	When running—brakes sticking; look back for hand signals.

17. The headlight will be displayed to the front of every train by night. It must be concealed or extinguished when a train turns out to meet another and has stopped clear of main track.

It must be dimmed:

(a) While passing through yards where yard engines are employed;

(b) Approaching stations at which stops are to be made or where trains are receiving or discharging passengers;

(c) Approaching train order signals, junctions, terminals, meeting points, or while standing on main track;

(d) On two or more tracks when approaching train in the opposite direction.

When an engine is running backward a white light must be displayed by night on the leading end.

FIG. 1.—Running backward by night, without cars or at the front of a train pulling cars. White light at A.

18. Yard engines will display the headlight to the front and rear by night. When not provided with a headlight at the rear, a white light must be displayed.

19. The following signals will be displayed to the rear of every train, as markers, to indicate the rear of the train:

A train not equipped to display the prescribed signals

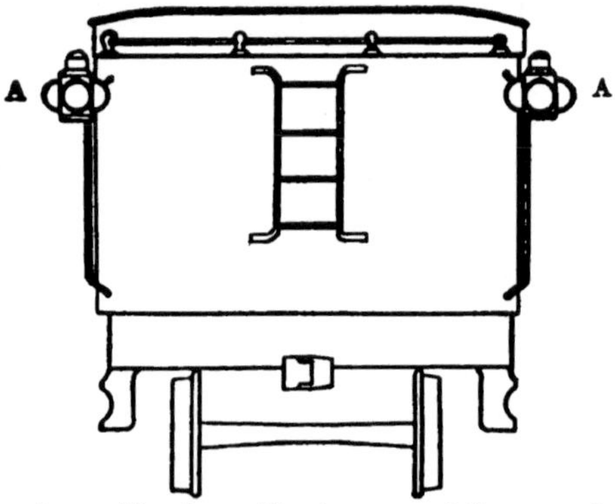

FIG. 2.—Running forward by day, without cars or at the rear of a train pushing cars. Marker lamps not lighted, or yellow (or green) flags at A A as markers.

NOTE TO RULE 18.—Headlight on end coupled to cars may be extinguished.

NOTE TO RULES 17, 19, 20 and 21.—The diagrams are intended to illustrate the general location of the train signals, not the exact manner in which they are to be attached.

will display a red flag by day and a red light by night to indicate the rear of the train, except when clear of the main track a white light instead of a red light will be displayed.

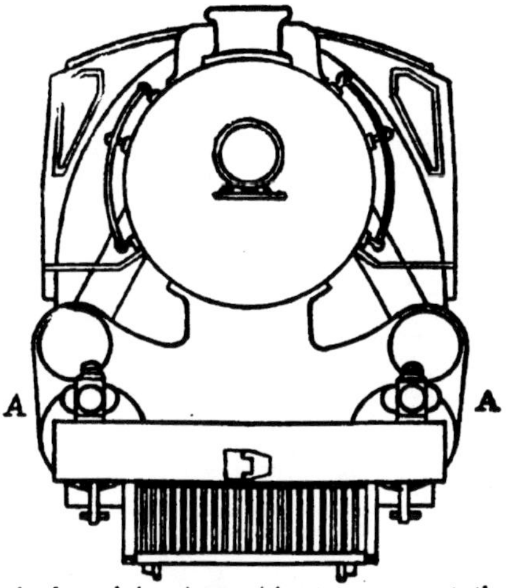

FIG. 3.—Running backward by day, without cars or at the rear of a train pushing cars. Marker lamps not lighted, or yellow (or green) flags at A A as markers.

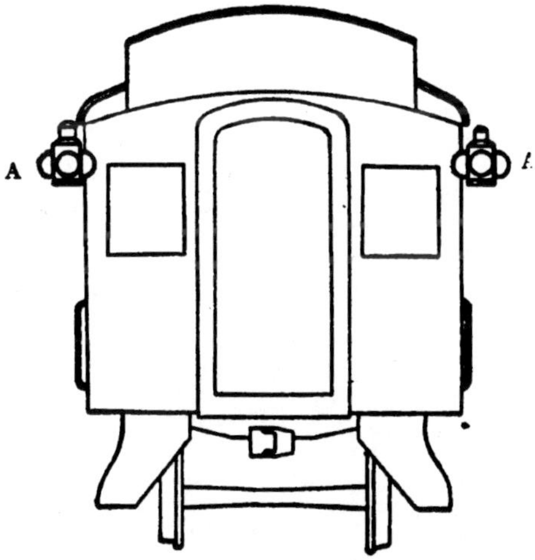

FIG. 4.—Rear of train by day. Marker lamps not lighted, or yellow (or green) flags at A A as markers.

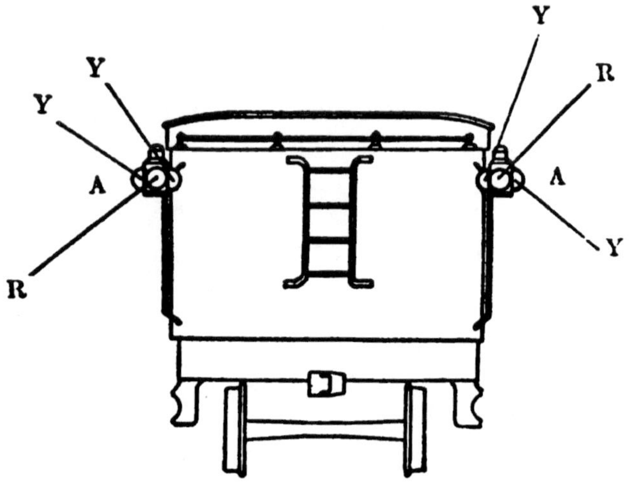

Fig. 5.—Running forward by night, without cars or at the rear of a train pushing cars, on single track, with the current of traffic on two tracks, with the current of traffic on passenger tracks where there are three or more tracks, and on tracks where trains move under the provisions of Rule 261. Lights at A A as markers, showing yellow (or green) to the front and side and red to the rear.

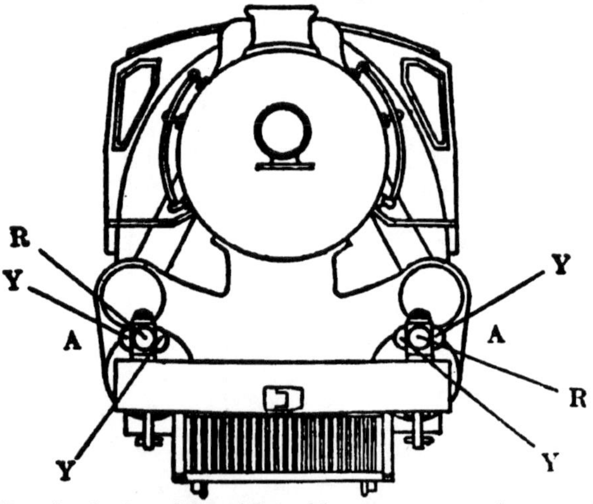

Fig. 6.—Running backward by night, without cars, or at the rear of a train pushing cars, on single track, with the current of traffic on two tracks, with the current of traffic on passenger tracks where there are three or more tracks and on tracks where trains move under the provisions of Rule 261. Lights at A A as markers, showing yellow (or green) to side and in direction engine is moving and red in opposite direction.

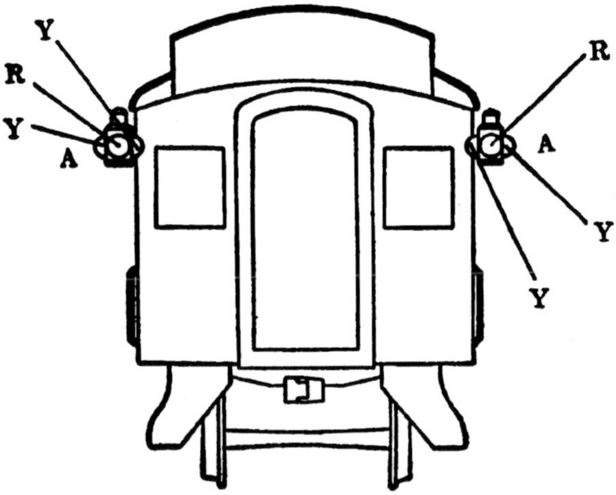

FIG. 7.—Rear of train by night while running on single track, with the current of traffic on two tracks, with the current of traffic on passenger tracks where there are three or more tracks, and on tracks where trains move under the provisions of Rule 261. Lights at A A as markers, showing yellow (or green) to the front and side and red to the rear.

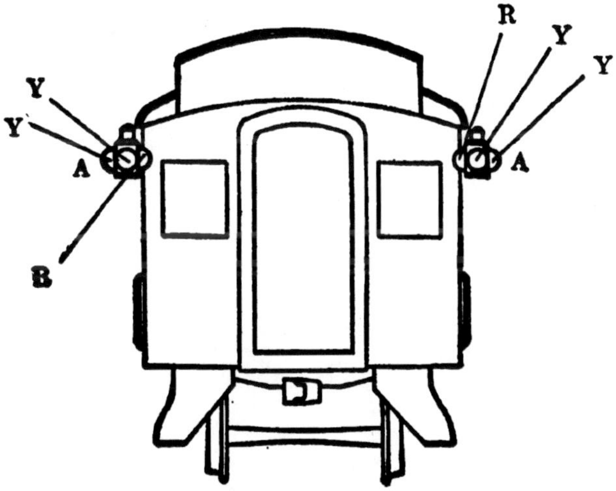

FIG. 8.—Rear of train by night when clear of main track. Lights at A A as markers, showing yellow (or green) to the front, side and to rear.

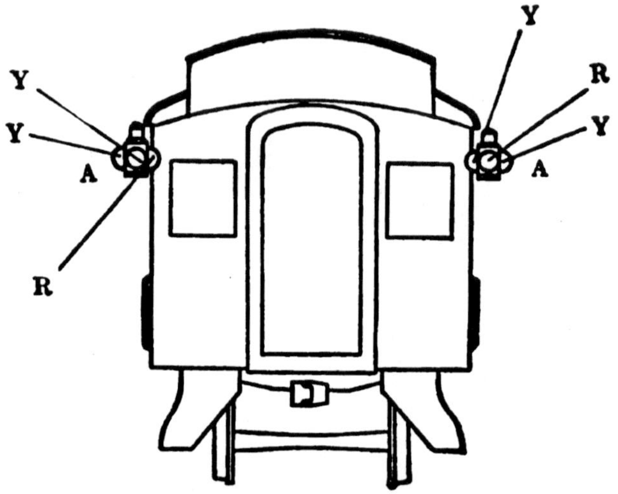

FIG. 9.—Rear of train by night running with the current of traffic on .... track where there are three or more tracks. Lights at A A as markers, showing yellow (or green) to rear on the side next to the .... track in the direction of current of traffic and red on the opposite side.

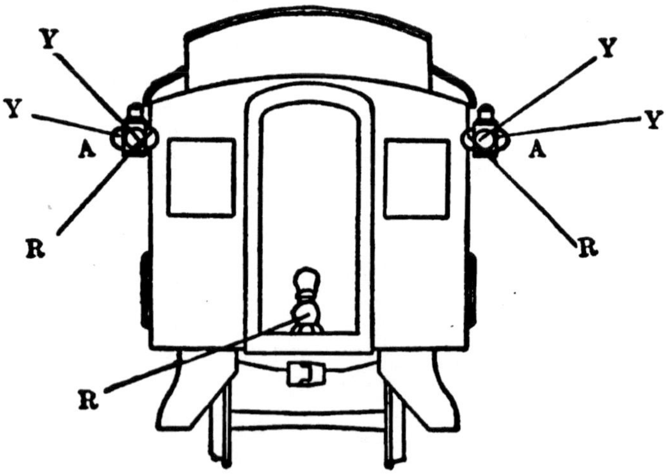

FIG. 10.—Rear of train by night running on any track against the current of traffic where there are three or more tracks. Lights at A A as markers, showing yellow (or green) to the rear, with a red light on the platform or the cupola.

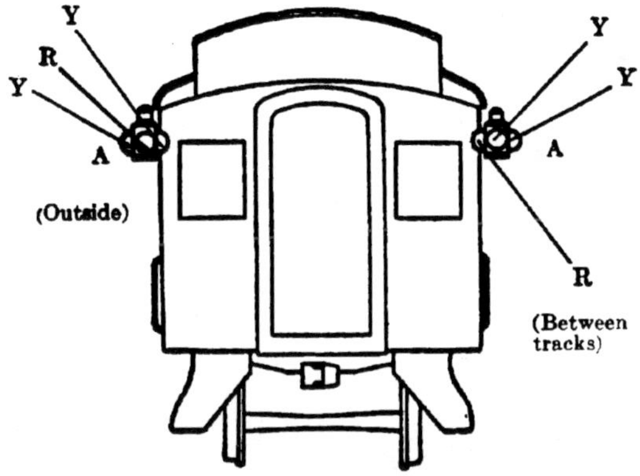

(This illustration is for a road which uses the right hand track.)

Fig. 11.—Rear of train by night running against the current of traffic on two tracks. Lights at A A as markers, showing yellow (or green) to front and side and yellow (or green) to the rear on the side next to the main track on which the current of traffic is in the direction the train is moving and red to the rear on the opposite side.

20. All sections except the last will display two green flags, and, in addition, two green lights by night in the places provided for that purpose on the front of the engine.

Note to Rules 19, 20 and 21.—When rear unit of a train is equipped with built-in electric markers, they must be lighted by day as well as by night, and the provision that markers will display yellow to the front may be omitted.

Note to Rules 20 and 21.—Where flags are prescribed, suitable lights may be substituted.

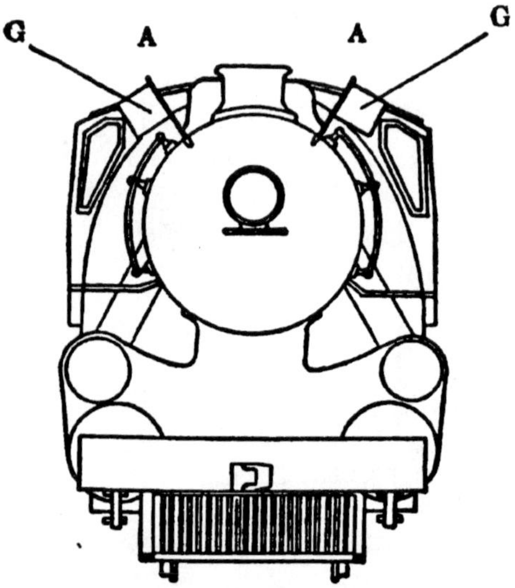

FIG. 12.—Running forward by day displaying signals for a following section. Green flags at A A.

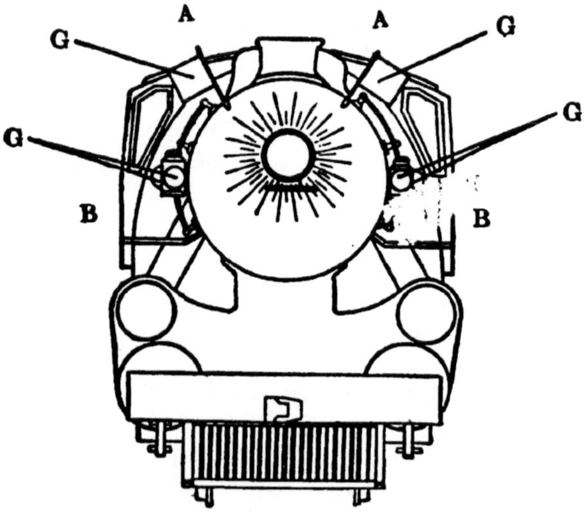

FIG. 13.—Running forward by night displaying signals for a following section. Green flags at A A and green lights at B B.

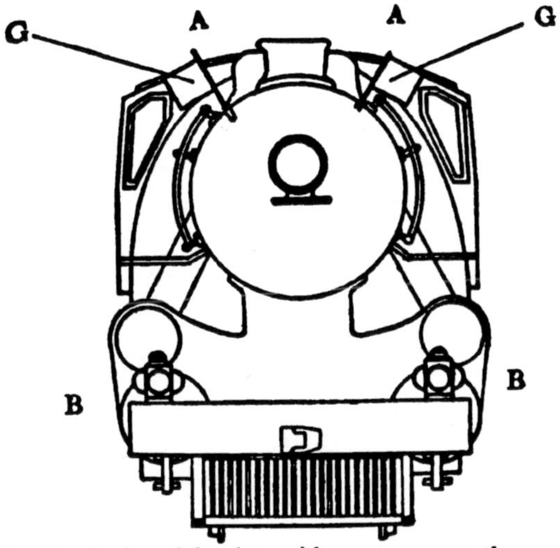

FIG. 14.—Running backward by day, without cars or at the rear of a train pushing cars, and displaying signals for a following section. Green flags at A A. Marker lamps not lighted, or yellow (or green) flags at B B as markers.

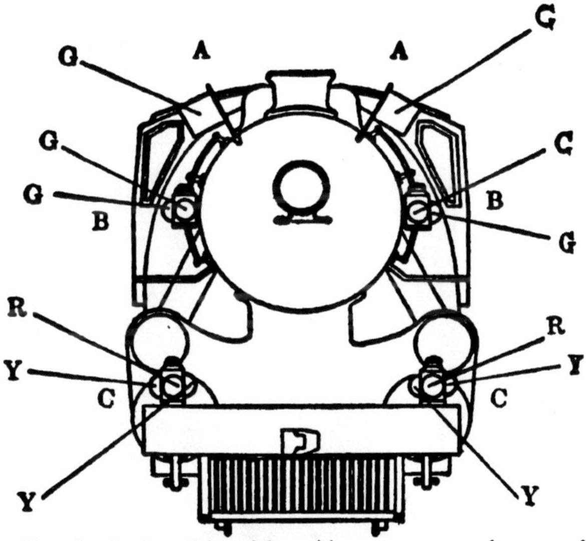

FIG. 15.—Running backward by night, without cars or at the rear of a train pushing cars, and displaying signals for a following section. Green flags at A A. Green lights at B B. Lights at C C as markers, showing yellow (or green) to side and in direction engine is moving and red in opposite direction.

21. Extra trains will display two white flags and in addition two white lights by night, in the places provided for that purpose on the front of the engine.

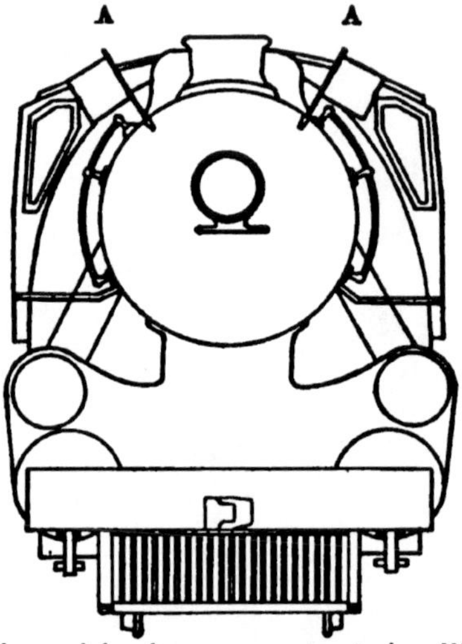

Fig. 16.—Running forward by day as an extra train. White flags at A A.

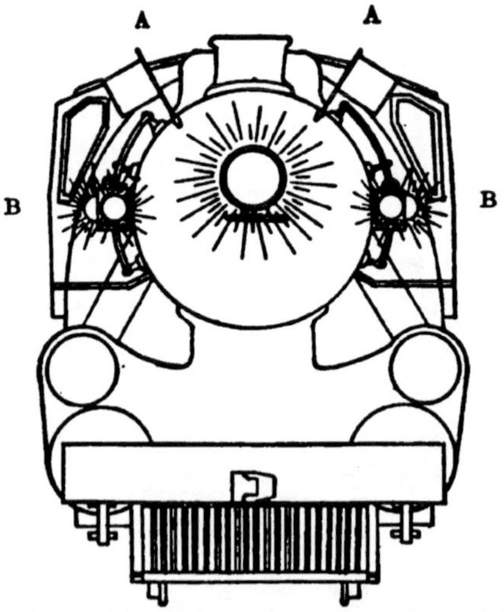

Fig. 17.—Running forward by night as an extra train. White flags at A A and white lights at B B.

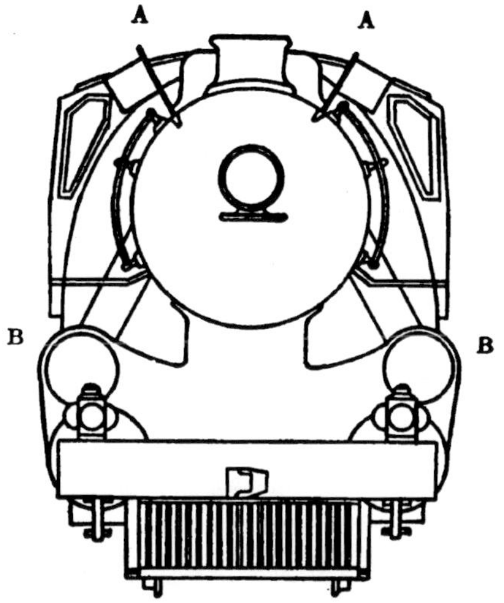

FIG. 18.—Running backward by day as an extra train, without cars or at the rear of a train pushing cars. White flags at A A. Marker lamps not lighted, or yellow (or green) flags at B B as markers.

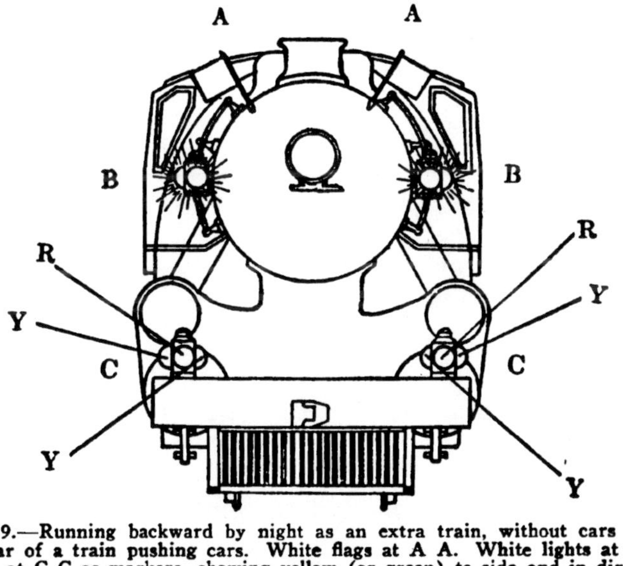

FIG. 19.—Running backward by night as an extra train, without cars or at the rear of a train pushing cars. White flags at A A. White lights at B B. Lights at C C as markers, showing yellow (or green) to side and in direction engine is moving and red in opposite direction.

22. When two or more engines are coupled, each engine shall display the signals as prescribed by Rules 20 and 21.

23. One flag or light displayed where in Rules 20 and 21 two are prescribed will indicate the same as two; but the proper display of all train signals is required.

24. When two or more engines are coupled, the leading engine shall sound the signals as prescribed by Rule 14.

25. Each car of a passenger train must be connected with the engine by a communicating signal appliance.

26. A blue signal, displayed at one or both ends of an engine, car or train, indicates that workmen are under or about it; when thus protected it must not be coupled to or moved. Each class of workmen will display the blue signals and the same workmen are alone authorized to remove them. Other equipment must not be placed on the same track so as to intercept the view of the blue signals, without first notifying the workmen.

When emergency repair work is to be done under or about cars in a train and a blue signal is not available, the engineman and fireman will be notified and protection must be given those engaged in making the repairs.

27. A signal imperfectly displayed, or the absence of a signal at a place where a signal is usually shown, must be regarded as the most restrictive indication that can be given by that signal, except that when the day indication is plainly seen, or when sufficient lights in a position or color position light signal are

---

Note to Rule 27.—Reflectors may be substituted for switch lights.

displayed to determine indication of the signal, it will govern.

Engine and train crews using a switch where the switch light is imperfectly displayed or absent, must, if practicable, correct or replace the light.

A signal imperfectly displayed or the absence of a signal at a place where a signal is usually shown, must be promptly reported to the ....

28. A white signal will be used to stop a train at the flag stations indicated on its schedule or in special instructions.

29. When a signal, except a fixed signal, is given to stop a train, it must, unless otherwise provided, be acknowledged as prescribed by Rule 14 (g) or (h).

30. The engine bell must be rung when an engine is about to move and while approaching and passing public crossing at grade and ....

32. The unnecessary use of either the whistle or the bell is prohibited.

33. Watchman stationed at public crossings at grade must use stop signals when necessary to stop trains. They will use prescribed signals to stop highway traffic.

34. All members of engine and train crews must, when practicable, communicate to each other by its name the indication of each signal affecting movement of their train or engine.

35. The following signals will be used by flagmen:

Day signals—A red flag,
Torpedoes and
Fusees.

Night signals—A red light,
Torpedoes and
Fusees.

SUPERIORITY OF TRAINS

S-71. A train is superior to another train by right, class or direction.

Right is conferred by train orders; class and direction by time-table.

Right is superior to class or direction.

Direction is superior as between trains of the same class.

D-71. A train is superior to another train by right or class.

Right is conferred by train order; class by time-table.

Right is superior to class.

72. Trains of the first class are superior to those of the second; trains of the second class are superior to those of the third; and so on.

S-72. Trains in the direction specified by the time-table are scperior to trains of the same class in the opposite direction.

73. Extra trains are inferior to regular trains.

MOVEMENT OF TRAINS AND ENGINES

82. Time-table schedules, unless fulfilled, are in effect for twelve hours after their time at each station.

Regular trains more than twelve hours behind either their schedule arriving or leaving time at any station lose both right and schedule, and can thereafter proceed only as authorized by train order.

S-83. A train must not leave its initial station on any division, or subdivision, or a junction, or pass from one of two or more tracks to single track, until it has been ascertained whether all trains due, which are superior, have arrived or left.

Stations at which train registers are located will be designated by time-table.

D-83. A train must not leave its initial station on any division, or subdivision, or a junction, until it has been ascertained whether all superior trains due have left.

Stations at which train registers are located will be designated by time-table.

84. A train must not start until the proper signal is given.

85. Trains of one schedule may pass trains of another schedule of the same class, and extra trains may pass and run ahead of ...... class trains and extra trains.

A section must not pass and run ahead of another section of the same schedule without first exchanging train orders, signals and numbers with the section to be passed. The change in sections must be reported from the next available point of communication.

86. Unless otherwise provided, an inferior train must be clear at the time a superior train in the same direction is due to leave the next station in the rear where time is shown.

S-87. An inferior train must keep out of the way of opposing superior trains and failing to clear the main track by the time required by rule must be protected as prescribed by Rule 99.

Extra trains must clear the time of opposing regular trains not less than five minutes unless otherwise provided, and will be governed by train orders with respect to opposing extra trains.

S-88. At meeting points between extra trains, the train

in the inferior time-table direction must take the siding unless otherwise provided.

Trains must pull into the siding when practicable; if necessary to back in, the train must first be protected as prescribed by Rule 99, unless otherwise provided.

S-89. At meeting points the inferior train must take the siding and clear the time of the superior train not less than five minutes, except at schedule meeting points between trains of the same class, where the inferior train must clear the main track before the leaving time of the superior train.

The superior train must stop at schedule meeting points with trains of the same class unless switch is properly lined and track clear. The inferior train must pull into the siding when practicable. If necessary to back in, unless otherwise provided it must be protected as prescribed by Rule 99.

S-90. The engineman of each train will give signal 14 (n) at least one mile before reaching a meeting or waiting point.

91. Unless some form of block system is used, trains in the same direction must keep not less than ten minutes apart, except in closing up at stations.

92. A train must not leave a station in advance of its schedule leaving time.

93. Within yard limits the main track may be used, clearing the time of first class trains at the next station where time is shown. Protection against .... class, extra trains and engines is not required.

Note to Rule 93.—Where block signal rules are in effect, "known to be clear" includes when track is known to be clear by signal indication.

. . . . class, extra trains and engines must move within yard limits at yard speed unless the main track is known to be clear.

A train or engine must not be moved against the current of traffic within yard limits until provision has been made for the protection of such movement.

94. A train which overtakes another train so disabled that it cannot proceed will pass it, if practicable, and if necessary will assume the schedule and take the train orders of the disabled train, proceed to the next available point of communication, and there report to . . . . The disabled train will assume the right or schedule and take the train orders of the last train with which it has exchanged and will, when able, proceed to and report from the next available point of communication.

95. Two or more sections may be run on the same schedule. Each section has equal time-table authority.

Unless otherwise provided, a train must not display signals for a following section without train order authority.

S-96. Unless otherwise provided, signals must not be ordered displayed to, nor taken down at, other than a register station for the train displaying the signals

97. Unless otherwise provided, extra trains must not be run without train orders.

98. Trains and engines must approach the end of two or more tracks, junctions, railroad crossings at grade, and drawbridges, prepared to stop, unless the switches are properly lined, signals indicate proceed, and track is clear. Where required by law, trains must stop.

99. When a train is moving under circumstances in which it may be overtaken by another train, the flag-

man must drop lighted fusees at proper intervals and take such other action as may be necessary to insure full protection.

When a train stops under circumstances in which it may be overtaken by another train, the flagman must go back immediately with flagman's signals a sufficient distance to insure full protection, placing two torpedoes and, when necessary, in addition, displaying lighted fusees. When recalled and safety to the train will permit, he may return, leaving the torpedoes and a lighted fusee.

When a train stops under circumstances in which it may be overtaken by another train, the engineman will immediately signal the flagman to protect the rear. When ready to proceed he will recall the flagman.

The front of the train must be protected in the same way when necessary by the ....

Conductors and enginemen are responsible for the protection of their trains.

101. Trains and engines must be fully protected against any known condition, not covered by the rules, which interferes with their safe passage.

102. When a train is disabled or stopped suddenly by an emergency application of the air brakes or other causes, adjacent tracks as well as tracks of other railroads that are liable to be obstructed must at once be protected until it is ascertained they are safe and clear for the movement of trains.

103. When cars are pushed by an engine, and the conditions require, a trainman must take a conspicuous

---

Note to Rule 99.—When trains are operating under Automatic Block System Rules, protection against following trains will have been complied with when full protectlon is afforded against trains moving at restricted speed.

position on the leading car and when shifting over public crossings at grade not protected by a watchman, or by gates, a member of the crew must protect the crossing.

104. Conductors are responsible for the position of switches used by them and members of their crews, except where switch tenders are stationed. Switches must be properly lined after having been used.

A switch must not be left open for a following train or engine unless in charge of a member of the crew of such train or engine.

Employes lining switches must see that points fit properly.

When practicable, the engineman must see that the switches near the engine are properly lined.

A train or engine must not foul a track until switches connected with the movement are properly lined, or in the case of spring switches the normal route is seen to be clear. When waiting to cross from one track to another and during the approach or passage of a train or engine on tracks involved, all switches connected with the movement must be secured in the normal position. Switches must not be restored to normal position until a movement is completed or clear of the main track involved.

Where trains or engines are required to be reported clear of main track such report must not be made until switch has been secured in its normal position.

105. Unless otherwise provided trains and engines using other than a main track must proceed at yard speed.

NOTE TO RULE 104.—Rule 104 applies only to hand operated switches.

When spring or remotely controlled switches are operated by hand, they are hand operated switches and Rule 104 applies.

Sidings of an assigned direction must not be used in a reverse direction unless authorized by the .... or in an emergency under flag protection.

106. Both the conductor and the engineman are responsible for the safety of the train and the observance of the rules, and, under conditions not provided for by the rules, must take every precaution for protection.

This does not relieve other employes of their responsibility under the rules.

107. When a passenger train is receiving or discharging traffic on the side toward a station, a train or engine must not pass between it and the station unless proper safeguards are provided.

108. In case of doubt or uncertainty, the safe course must be taken.

D-151. Where two main tracks are in service trains must keep to the .... unless otherwise provided.

Where three or more main tracks are in service they shall be designated by numbers and their use indicated by special instructions.

D-152. When a train or engine crosses over to, or obstructs another track, unless otherwise provided, it must first be protected as prescribed by Rule 99.

153. When a trailing movement through a spring switch is stopped before passing entirely through the switch, the movement must not be reversed until the switch has been properly set by hand.

Trains and engines must stop and examine main track spring switches before making facing point movements over them, unless receiving a signal indication permitting them to proceed.

RULES FOR MOVEMENT BY TRAIN ORDERS

201. For movements not provided for by time-table, unless otherwise provided, train orders will be issued by authority and over the signature of the *(Dispatcher or Chief Dispatcher)* and only contain information or instructions essential to such movements.

They must be brief and clear; in the prescribed forms when applicable; and without erasure, alteration or interlineation.

Figures in train orders must not be surrounded by brackets, circles or other characters.

202. Each train order must be given in the same words to all employes or trains addressed.

203. Train orders must be numbered consecutively each day, beginning at midnight.

204. Train orders must be addressed to those who are to execute them, naming the place at which each is to receive his copy. Those for a train must be addressed to the conductor and engineman, and also to anyone who acts as its pilot. A copy for each employe addressed must be supplied by the operator. Orders addressed to operators restricting the movement of trains must be respected by conductors and enginemen the same as if addressed to them.

205. Each train order must be written in full in a book provided for the purpose in the office of the train dispatcher; and with it recorded the time and the signals which show when, from what offices and by whom the order was repeated and the responses transmitted; and the train dispatcher's initials. These records must be made at once and never from memory or memoranda.

Additions to train orders must not be made after they have been repeated.

206. In train orders regular trains will be designated by numbers as "No. 10," and sections as "Second 10," adding engine numbers if desired. Extras will be designated by engine numbers and the direction, as "Extra 798, 'East' or 'West'." Work extras will be designated by engine numbers, as "Work extra 798." For the movement of an engine of another company, the initials will precede the engine number.

When two or more engines are coupled and a designation is made by engine numbers, the number of the leading engine will be used in train orders.

Even hours as "10 A.M." must not be used in stating time in train orders.

In transmitting train orders time may be stated in figures only or duplicated in words.

In transmitting train orders by telephone, the numbers of trains and engines in the address may be pronounced and then spelled, letter by letter if so desired. All stations and numerals in the body of an order must first be plainly pronounced and then spelled, letter by letter thus: Aurora, A-u-r-o-r-a, and One Naught Five, O-n-e N-a-u-g-h-t F-i-v-e.

The letters duplicating names of stations and numerals will not be written in the order book nor upon train orders, except time, which may be duplicated in words.

When train orders are transmitted by telegraph, the train dispatcher must write the order as repeated by the first operator and must underscore each word and figure in the order as it is repeated by other operators. When transmitted by telephone, he must write the order as he transmits it and underscore each word and figure as repeated by each operator.

207. To transmit a train order, the signal "31" or "19" followed by the direction must be given to each office addressed, stating the number of copies—thus, "31, West, copy 5" or "19, East, copy 2".

208. A train order to be sent to two or more offices must be transmitted simultaneously to as many of them as practicable. When not sent simultaneously to all, the order must be sent first to the superior train.

The several addresses must be in the order of superiority of trains, each office taking its proper address, and, when practicable, must include the operator at the meeting or waiting point.

Copies of the order addressed to the operator at the meeting or waiting point must be delivered to the trains affected until all have arrived from one direction.

A train order restricting the movement of a train must not be issued for it at the point where such movement is restricted if it can be avoided. When so sent, the fact will be stated in the order, the train will be brought to a stop before delivery is made, and special precautions must be taken to insure safety.

209. Operators receiving train orders must write or typewrite them in manifold during transmission. If the requisite number of copies cannot be made at one writing, they must make others from the original copy and repeat to the train dispatcher from the new copies each time additional copies are made. They must retain a copy of each train order.

The word "complete," the time, and the signature of the operator must be in his handwriting.

210. When a "19" train order has been transmitted, operators must, unless otherwise directed, repeat it at once from the manifold copy, in the succession in which the several offices have been addressed. Each operator receiving the order should observe whether the others repeat correctly. When the order has been repeated correctly by an operator, the response "Complete" and the time, with the initials of the .... will be given by the train dispatcher. The operator receiving this response will then write in the space provided the word "Complete," the time and his last name in full, and deliver a copy to each person addressed. But when delivery to engineman will take the operator from the immediate vicinity of his office, the engineman's copy will be delivered by ....

Enginemen must show train orders to firemen and when practicable to forward trainmen. Conductors must show train orders when practicable to trainmen. Firemen and trainmen are required to read them and, if necessary, remind enginemen and conductors of their contents.

210. (Optional) When a "31" train order has been transmitted, operators must, unless otherwise directed, repeat it at once from the manifold copy in the succession in which the several offices have been addressed, and then write the time of repetition on the order. Each operator receiving the order should observe whether the others repeat correctly.

The conductor or engineman and others addressed must read it to the operator and then sign it, and the operator will send their signatures preceded by the number of the order to the . . . . The response "Complete," and the time, with the initials of the . . . . , will then be given by the train dispatcher. Each operator receiving this response will then write on each copy the word "Complete," the time, and his last name in full, and then deliver a copy to each person addressed, except that when the order is signed by the conductor, he must personally deliver a copy of it to each engineman; the engineman will then read the order to the conductor before proceeding.

Enginemen must show train orders to firemen and when practicable to forward trainmen. Conductors must show train orders when practicable to trainmen. Firemen and trainmen are required to read them and, if necessary, remind enginemen and conductors of their contents.

11. Clearance Form A must be filled out by the operator before clearing a train, showing thereon, without erasure or alteration, the total number of train orders and the number of each train order, if any, addressed to a train. He will then repeat from Clearance Form A to the train dispatcher the information shown thereon. The dispatcher will make the required record in the train order book, and if operator has correctly repeated the numbers of all train orders addressed to a train will respond by giving "OK," the time and his initials, which the operator will endorse on the Clearance Form A.

If all orders held for a train have been completed, and communication fails before Clearance Form A has been OK'd by train dispatcher, the operator may deliver such orders accompanied by Clearance Form A, and bearing notation "Wire has failed," which will be acted upon as though "OK" had been given in the usual way. When communication is restored, operator will notify train dispatcher of each train cleared, the time, and the numbers of the orders delivered, which train dispatcher will record in train order book.

Clearance Form A must be delivered together with all train orders to each person addressed. Conductors, enginemen and firemen must, and when practicable trainmen will, see that the information shown on Clearance Form A corresponds with the train orders received.

Operators will retain a copy of each Clearance Form A.

212. When so directed by the train dispatcher, a train order, having been transmitted, may be acknowledged before repeating, by the operator responding:

"Order Number .... to Number (of train) ...., X," with the operator's initials and office signal. The operator must then write on the order his initials and the time.

213. "Complete" must not be given to a train order for delivery to an inferior train until the order has been repeated or the "X" response sent by the operator who receives the order for the superior train.

214. When a train order has been repeated or "X" response sent, and before "Complete" has been given, the order must be treated as a holding order for the train addressed, but must not be otherwise acted on until "Complete" has been given.

If the means of communication fails before an office has repeated an order or has sent the "X" response, the order at that office is of no effect and must be there treated as if it has not been sent.

216. When the train dispatcher issues a train order that

is to be delivered to a train by the dispatcher, a carbon impression must be made in the train order book at the time the order is written. If later the order is to be sent to another office, it will be transmitted from the copy in the book. The requirements for delivery are the same as at other offices.

?17. A train order to be delivered to a train at a point not a train order office, or at which the office is closed, must be addressed to "C & E .... at (or between) ...., care of ....," and forwarded and delivered by the conductor or other person in whose care it is addressed, and who is responsible for its delivery. The numbers of such train orders must be shown in the usual manner on Clearance Form A of the train making delivery.

For orders which are sent in the manner herein provided, to a train, the superiority of which is thereby restricted, the operator will be directed to make an extra copy of the order, which he will deliver to the person who is to make delivery of the order. On this copy, the person delivering the order must secure the signatures of the conductor and engineman addressed. This copy he must deliver to the first operator accessible, who must at once transmit the signatures of the conductor and engineman to the train dispatcher, and preserve the copy. Under such circumstances "Complete" must not be given to the order for an inferior train until the train dispatcher has received the signatures of the conductor and engineman of the superior train.

When form "31" is used, "Complete" will be given upon the signature of the person by whom the order is to be delivered.

?18. When a train is named in a train order by its schedule number alone, all sections of that schedule are included, and each must have copies delivered to it.

219. An operator must not repeat or give the "X" response to a train order restricting the movement of a train which has been cleared, or of which the engine has passed his train order signal in proceed position, until he has obtained the signatures of the conductor and engineman to the order.

If necessary to issue a second Clearance Form A, the original Clearance Form A must be destroyed, and the second Clearance Form A must show the numbers of all train orders addressed to the train.

220. Train orders once in effect continue so until fulfilled, superseded or annulled. Any part of an order specifying a particular movement may be either superseded or annulled.

Orders held by or issued for or any part of an order relating to a regular train become void when such train loses both right and schedule as prescribed by Rules 4 and 82, or its schedule is annulled.

When a conductor or engineman, or both, is relieved before the completion of a trip, all train orders and instructions held must be delivered to the relieving conductor or engineman. Such orders or instructions must be compared by the conductor and engineman before proceeding.

221. Unless otherwise provided, a fixed signal must be used at each train-order office, which shall indicate "stop" when there is an operator on duty, except when changed to "proceed" to allow a train to pass when there are no train orders for any train in that direction. While "stop" is indicated, trains must not leave without Clearance Form A.

222. Operators must promptly record and report to the .... the time of arrival and departure of all trains and the direction of extra trains.

They must observe trains and report at once to the .... if the proper signals are not displayed.

223. The following signals and abbreviations may be used:

Initials for signature of ....

Such office and other signals as are arranged by the ....

C & E	–for Conductor and Engineman.
C & –	–for Conductor and ....
X	–Train will be held until train order is made "Complete."
Com	–for Complete.
OK	–for Correct.
O S	–Train Report.
No	–for Number.
Eng	–for Engine.
Sec	–for Section.
Psgr	–for Passenger.
Frt	–for Freight.
Mins	–for Minutes.
Jct	–for Junction.
MP	–for Mile Post.
MPH	–for Miles Per Hour.
Dispr	–for Train Dispatcher.
Opr	–for Operator.
31 or 19	–to clear the line for Train Order, and for Operators to ask for Train Orders.

The usual abbreviations for the names of the months and stations.

FORMS OF TRAIN ORDERS

FORM S-A

FIXING MEETING POINTS FOR OPPOSING TRAINS

(1) No. 1 meet No. 2 at B.
No. 3 meet Second 4 at B.
No. 5 meet Extra 95 east at B.
Passenger Extra 652 north meet Extra 231 south at B.

(2) No. 2 and Second 4 meet No. 1 and No. 3 at C and Extra 95 west at D (and so on).

No. 1 meet No. 2 at B Second 4 at C and Extra 95 east at D.

Trains receiving these orders will run with respect to each other to the designated points and there meet in the manner prescribed by the rules.

FORM B

DIRECTING A TRAIN TO PASS OR RUN AHEAD OF ANOTHER TRAIN

(1) Extra 594 west pass No. 1 at J.

Both trains will run according to rule to the designated point and there arrange for the rear train to pass promptly. Unless otherwise provided, the train to be passed will take siding.

When an inferior train receives an order to pass a superior train, authority is conferred to run ahead of the train passed from the designated point.

(2) Extra 594 east run ahead of No. 6 M to B.

The first-named train will run ahead of the second-named train between the designated points.

FORM S-C

GIVING RIGHT OVER AN OPPOSING TRAIN

(1) No. 1 has right over No. 2 G to X.

If the second-named train reaches the point last named before the other arrives, it may proceed, keeping clear of the schedule of opposing train as required by rule.

(2) Extra 37 east has right over No. 3 F to A.

The regular train must not go beyond the point last named until the extra train has arrived, unless authorized by train order to do so.

These orders give right to the train first-named over the other train between the points named. If the trains meet at either of the designated points, the first-named train must take the siding, unless the order otherwise prescribes.

(3) Extra 72 south has right over Extra 91 north A to Z and wait at

N until 2 10 p m

P 3 25 p m for Extra 91 North.

The first-named extra train must not pass the designated waiting points before the time given, unless the second-named extra train has arrived. The second-named extra train must clear the time specified at the designated points or any intermediate station not less than 5 minutes.

FORM D

GIVING RIGHT OVER ANOTHER TRAIN IN THE SAME DIRECTION

(1) No. 1 has right over No. 3 A to Z.

The second-named train must clear the time of the first-named train, between the designated points, as required by rule.

(2) Extra 62 east has right over No. 4 Z to A and wait at:
Z until 12 50 p m
Y 1 10 p m
X 1 20 p m

(3) Extra 57 west has right over Extra 61 west A to Z.

These orders give right to the first-named train over the other train between the points named. If train order time is given, the first-named train must not pass the designated stations before the time shown, and the second-named train must clear such times as required by rule. If no time is given, the second-named train must not leave the point first named ahead of the first-named train.

FORM E — TIME ORDERS

(1) No. 1 run 50 mins late A to G.

No. 1 run 50 mins late A to G and 20 mins late G to K, etc.

This makes the schedule time of the train named, between the stations designated, as much later as stated in the order, and any other train receiving the order is required to run with respect to this later time, as before required to run with respect to the regular schedule time. The time in the order should be such as can be easily added to the schedule time.

(2) No. 1 and No. 3 wait at N until 9 59 a m
P 10 30 a m
R 10 55 a m, etc.

The train, or trains, named must not pass the designated points before the times given. Other trains receiving the order are required to run with respect to the time specified at the designated points or any intermediate station where schedule time is earlier than the time specified in the order, as before required to run with respect to the schedule time of the train, or trains, named.

FORM S-E

(1) No. 1 wait at H until 9 59 a m for No. 2.

The train first-named must not pass the designated point before the time given, unless the other train has arrived. The train last-named is required to run with respect to the time specified, at the designated point or any intermediate station where schedule time is earlier than the time specified in the order, as before required to run with respect to the schedule time of the train first-named.

FORM F — FOR SECTIONS

(1) Eng 20 display signals and run as First 1 A to Z.

To be used when the number of the engine for which signals are displayed is unknown, and is to be followed by (2), both being single-order examples.

(2) Eng 25 run as Second 1 A to Z.

(3) Second 1 display signals B to E for Eng 99.

(4) Engs 20, 25 and 99 run as First Second and Third 1 A to Z.

To add an intermediate section, (5) will be used.

(5) Eng 85 display signals and run as Second 1 A to Z. Following sections change numbers accordingly.

The engine named will display signals and run as directed, and following sections will take the next higher number.

To drop an intermediate section, (6) will be used.

(6) Eng 85 is withdrawn as Second 1 at H. Following sections change numbers accordingly.

The engine named will drop out at H, and following sections will take the next lower number.

To substitute one engine for another on a section, (7) will be used.

(7) Eng 18 instead of Eng 85 display signals and run as second 1 R to Z.

The second-named engine will drop out at R, and be replaced by the first-named engine.

Following sections need not be addressed.

If the second-named engine is the last section, the words "display signals and" will be omitted.

To discontinue the display of signals, (8) will be used.

(8) Second 1 take down signals at D.

The train named will take down signals as directed, and a following section must not proceed beyond the designated point.

To pass one section by another, (9) will be used.

(9) Engs 99 and 25 reverse positions as Second and Third 1 H to Z.

Conductors and enginemen of the trains addressed will exchange orders and signals. Following sections, if any, need not be addressed.

Each section affected by these orders must have copies and must arrange signals accordingly.

To annul a section for which signals have been displayed over a division, or any part thereof, when no train is to follow the signals, Form K must be used.

When sections are run to an intermediate point of a

schedule, the train orders must specify which engine or engines shall assume the schedule beyond such point.

FORM G — EXTRA TRAINS

(1) Eng 99 run extra A to F.

(2) Eng 99 run Passenger Extra A to F.

(3) Eng 99 run extra A to F and return to C.

The extra must go to F before returning to C.

FORM S-H — WORK EXTRA

(1) Eng 292 works extra 6 45 a m until 5 45 p m between D and E.

The work extra must, whether standing or moving, protect itself against extra trains within the working limits in both directions as prescribed by the rules. The time of regular trains must be cleared.

This may be modified by adding:

(a) Not protecting against eastward extra trains.

The work extra will protect only against westward extra trains. The time of regular trains must be cleared.

(b) Not protecting against extra trains.

Protection against extra trains is not required. The time of regular trains must be cleared.

(c) Extra 173 east wait at E until 9 15 a m.

Extra 209 east wait at E until 12 10 p m.

Other eastward extra trains wait at E until 5 45 p m for Work Extra 292.

The work extra will protect against the trains named after the times specified in the order.

The work extra must protect against westward extra trains.

The time of regular trains must be cleared.

When a work extra has been instructed by order to not protect against extra trains, and, afterward, it is desired to have it clear the track for a designated extra, an order may be given in the following form:

(2) Work extra 292 clears (or protects against) Extra 76 east between D and E after 2 10 p m.

Extra 76 east must not enter the working limits before 2 10 p m, and will then run expecting to find the work extra clear of the main track, or protecting itself, as the order may require.

To enable a work extra to work upon the time of a regular train, the following form will be used:

(3) Work Extra 292 protects against No 55 (or .... class trains) between D and E.

The work extra may work upon the time of the train or trains mentioned in the order, and must protect itself against such train or trains. The regular train or trains receiving the order will run expecting to find the work extra protecting itself.

When a work extra is to be given exclusive right over all trains, the following form will be used:

(4) Work Extra 292 has right over all trains between D and E 7 15 p m until 1 15 a m.

This gives the work extra the exclusive right between the points designated between the times named.

The working limits should be as short as practicable to be changed as the progress of the work may require.

Work extras must give way to all trains as promptly as practicable.

Whenever extra trains are run over working limits they must be given a copy of the order sent to the work extra. Should the working order instruct a work extra to not protect against extra trains in one or both directions, extra trains must protect against the work extra; if the order indicates that the work extra is protecting itself against other trains, they will run expecting to find the work extra protecting itself.

FORM D-H — WORK EXTRA

(1) Eng 292 works extra on (*Eastward*) track, or (*both*) tracks, 6 45 a m until 5 45 p m betwen D and E.

The work extra must, whether standing or moving, protect itself within the working limits against extra trains moving with the current of traffic on the track, or tracks named. The time of regular trains must be cleared.

This form may be modified by adding:

(a) Not protecting against extra trains.

Protection again extra trains is not required. The time of regular trains must be cleared.

(b) Extra 173 east wait at E until 9 15 a m.

Extra 209 east wait at E until 12 10 p m.

Other eastward extra trains wait at E until 5 45 p m.

The work extra will protect against the trains named after the times specified in the order. The time of regular trains must be cleared.

To enable a work extra to work upon the time of a regular train, the following form may be used:

(2) Work extra 292 protects against No. 55 between D and E.

The work extra may work upon the time of the train or trains mentioned in the order and must protect against such train or trains.

The regular train or trains receiving the order will run expecting to find the work extra protecting itself.

When it is desired to move a train against the current of traffic over the working limits, provisions must be made for the protection of such movement.

When a work extra is to be given exclusive right over all trains, the following form will be used:

(3) Work extra 292 has right over all trains on (*Eastward*) and (*Westward*) tracks between G and H 7 01 p m until 1 01 a m.

This gives the work extra the exclusive right to the track, or tracks, mentioned between the points designated between the times named.

The working limits should be as short as practicable, to be changed as the progress of the work may require.

Work extras must give way to all trains as promptly as practicable.

FORM J — HOLDING ORDER

(1) Hold No. 2.

(2) Hold all, or eastward, trains.

When a train has been so held it must not proceed until the order to hold is annulled, or an order given to the operator in the form:

.... may go.

These orders will be addressed to the operator and acknowledged in the usual manner, and will be delivered to conductors and enginemen of all trains affected.

Form J will be used only when necessary to hold trains until orders can be given, or in case of emergency.

FORM K — ANNULLING A SCHEDULE OR A SECTION

(1) No. 1 due to leave A Feb. 29th is annulled A to Z.

(2) Second 5 due to leave E Feb. 29th is annulled E to G.

The schedule or section annulled becomes void between the points designated and cannot be restored.

Form K will not be combined with other forms of train orders.

FORM L — ANNULLING AN ORDER

(1) Order No. 10 is annulled.

If an order which is to be annulled has not been delivered to a train, the annulling order will be addressed to the operator, who will destroy all copies of the order annulled but his own, and write on that: Annulled by Order No. ....

An order which has been annulled must not be reissued under its original number.

FORM S-M — ANNULLING PART OF AN ORDER

(1) That part of Order No. 10 reading Extra 263 west pass No. 1 at S is annulled.

(2) That part of Order No. 12 reading No. 3 pass No. 1 at S is annulled.

Form S-M will be used only when that part of the order not annulled is clear in its wording.

FORM S-P
SUPERSEDING AN ORDER OR A PART OF AN ORDER

This order will be given by adding to prescribed forms the words "instead of ...."

(1) No. 1 meet No. 2 at C instead of B.

An order which has been superseded must not be reissued under its original number.

When a train is directed by train order to take siding for another train, such instructions apply only at the point designated in that order, and do not apply to the superseding order unless so specified.

FORM P
SUPERSEDING AN ORDER OR PART OF AN ORDER

This order will be given by adding to prescribed forms, the words "instead of ...."

(1) No 1 pass No 3 at C instead of B.

An order which has been superseded must not be reissued under its original number.

FORM Q
ISSUANCE OF A NEW TIME-TABLE

(1) Time-table No. 1 is effective at 12 01 a m Sunday July 10.

Form Q will not be combined with other forms of train orders.

FORM D-R
PROVIDING FOR A MOVEMENT AGAINST THE CURRENT OF TRAFFIC

(1) No. 1 has right over opposing trains on (*Eastward*) track C to F.

The designated train must use the track specified between the points named and has right over opposing trains on that track between those points. Opposing trains must not leave the point last named until the designated train arrives.

All trains between the points named moving with the current of traffic in the same direction as the designated train must, when practicable, receive a copy of the order, and may then proceed on their schedules or rights.

This order may be modified as follows:

(2) After No. 4 arrives at C, No. 1 has right over opposing trains on (*Eastward*) track C to F.

The train to be moved against the current of traffic must not leave the first-named point until the arrival of the first-named train.

A train must not be moved against the current of traffic until the track on which it is to run has been cleared of opposing trains.

FORMS OF BLANKS FOR SINGLE AND TWO OR MORE TRACKS

SPECIFICATIONS FOR TRAIN ORDER FORM AND BOOKS FOR 31 ORDERS.

Form as here shown. Blank space for order 4". The mode of filling the blanks is indicated by small type.

Size 7" x 7½"; yellow translucent train order tissue, sized and of such thickness as to admit of making nine good manifold copies with stylus and double-faced carbon; printed black ink one side; 100 sheets to a pad;

printed manila top sheet; strawboard back glued at top with muslin strip.

Standard Train Order Blank for 31 Order.

FORM 31 — FORM 31

Name .... COMPANY

TRAIN ORDER No. 10

January 17 19 28

TO.............. **At**..............

X ....Initials.... **Opr.;** ....1 45 A **M.**

....

Each employee addressed must have a copy of this order.

Repeated at 2 20 A **M.**

Conductor or Engineman	Train	Made	Time	Operator
Jones Brown	45	Complete	2.20 a.m.	Black

Standard Train Order Blank for 19 Order.

FORM 19 FORM 19

Name **COMPANY**

TRAIN ORDER No. 10

January 17 **19** 28

To.............. **At**..............

X Initials **Opr.;** 1 45 A**M.**

Each employee addressed must have a copy of this order.

Made Complete **time** 2 16 P**M.** Black **Opr.**

(To be printed on white paper 5" x 6¾")

A **EAST AND WEST RAILROAD COMPANY** **A**

CLEARANCE FORM A

....19........

To C & E..at..

I have .... orders for your train.

(If no train orders, operator must write "No" in space provided above)

Orders:	No....................	No....................	No....................	No....................	No....................
	No....................	No....................	No....................	No....................	No....................
	No....................	No....................	No....................	No....................	No....................

*** When this form is used as authority for a train to pass Manual Block Signal in Stop position, condition of block will be shown below.**

Block..

Time OK..............................M. ....Dispr.

..............................Opr.

Manifold copies will be made for each Conductor, Engineman and Operator, the latter retaining a copy.

*** To be used only in connection with Rule 308.**

CHAPTER III

BLOCK SIGNAL RULES

RULES GOVERNING THE MOVEMENT OF TRAINS IN THE SAME DIRECTION BY BLOCK SIGNALS

251. On portions of the railroad, and on designated tracks so specified on the time-table, trains will run with reference to other trains in the same direction by block signals whose indications will supersede the superiority of trains.

253. The .... must be advised in advance, of any known condition that will delay the train or prevent it from making usual speed.

254. Except as affected by Rule 251 all Block Signal Rules and Operating Rules remain in force.

RULES GOVERNING OPPOSING AND FOLLOWING MOVEMENT OF TRAINS BY BLOCK SIGNALS

Note: Railroads operating under these rules must provide proper signals to control the approach and movement of trains.

261. On portions of the railroad, and on designated tracks so specified on the time-table, trains will be governed by block signals whose indications will supersede the superiority of trains for both opposing and following movements on the same track.

263. The signalman must be advised in advance of any known condition that will delay the train or prevent it from making usual speed.

264. Except as affected by Rule 261 all Block Signal Rules and Operating Rules remain in force.

FIXED SIGNALS

Rules 281 to 292, inclusive.

Aspects may be shown by the position of semaphore arms, color of lights, position of lights, flashing of lights, or a combination of color, position, and flashing of lights.

Day and night aspects for color light signals shall have the same colors as the night aspects of the semaphore signals.

Day and night aspects for position light signals shall

RULE 281

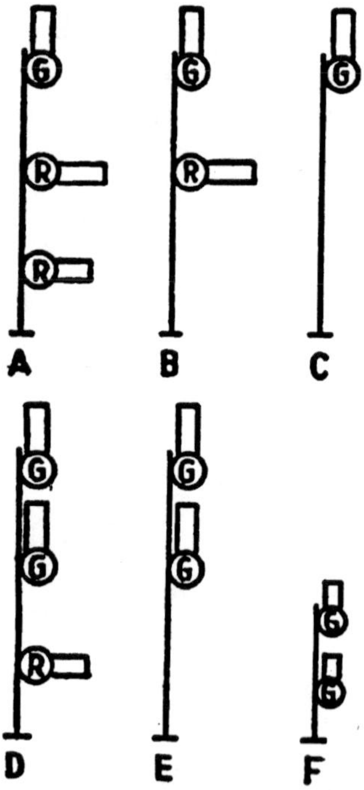

Indication – Proceed.
Name: Clear.

RULE 281A

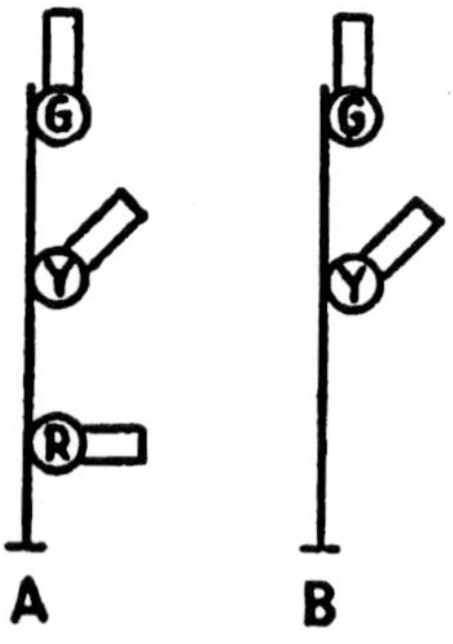

Indication—Proceed approaching second signal at medium speed.
Name: Advance approach medium.

have the same positions as the day aspects of the semaphore signals.

Aspects shown are typical. Each road should show the aspects and colors of lights it uses.

NOTE: In the following illustrations of typical signal aspects, Rules 281 to 292, inclusive,

R = Red
Y = Yellow
G = Green

NOTE: When flashing color lights are used, they shall be indicated as follows:

FR = Flashing Red
FY = Flashing Yellow
FG = Flashing Green

RULE 281B

Indication—Proceed approaching next signal at limited speed.

Name: Approach limited.

RULE 281C

Indication—Proceed; limited speed within interlocking limits.

Name: Limited—clear.

RULE 282

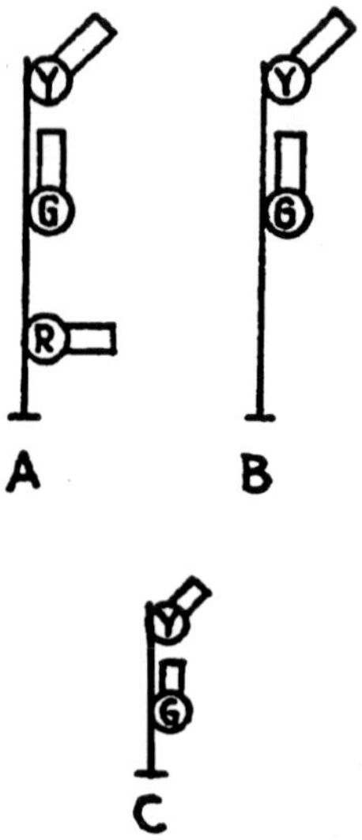

Indication—Proceed approaching next signal at medium speed.

Name: Approach medium.

RULE 283

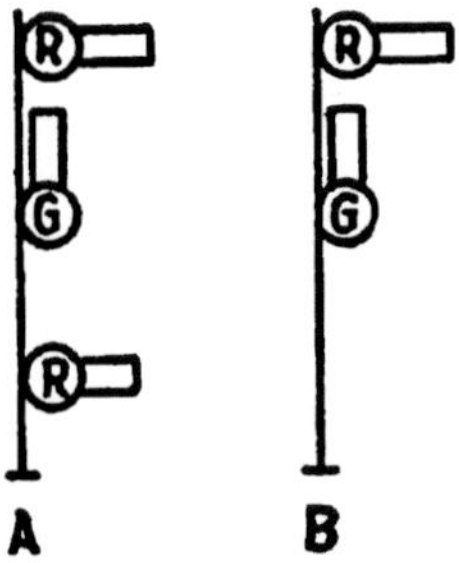

Indication — Proceed; medium speed within interlocking limits.

Name: Medium-clear.

RULE 282A

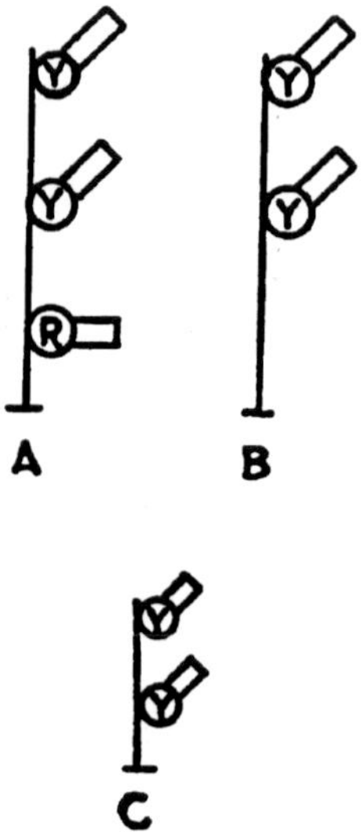

Indication—Proceed preparing to stop at second signal.

Name: Advance approach.

RULE 283A

Indication—Proceed preparing to stop at second signal; medium speed within interlocking limits.

Name: Medium — advance approach.

RULE 283B

Indication — Proceed at medium speed approaching next signal at slow speed.

Name: Medium—approach slow.

RULE 284

Indication—Proceed approaching next signal at slow speed. Train exceeding medium speed must at once reduce to that speed.

Name: Approach slow.

RULE 285

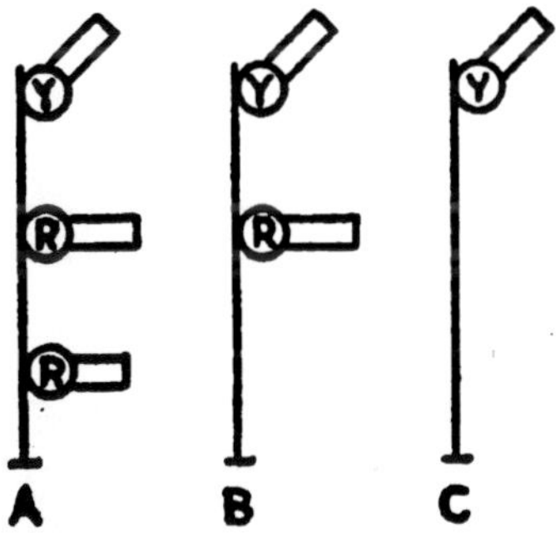

Indication—Proceed preparing to stop at next signal. Train exceeding medium speed must at once reduce to that speed.

Name: Approach.

RULE 286

Indication — Proceed at medium speed preparing to stop at next signal.

Name: Medium—approach.

RULE 287

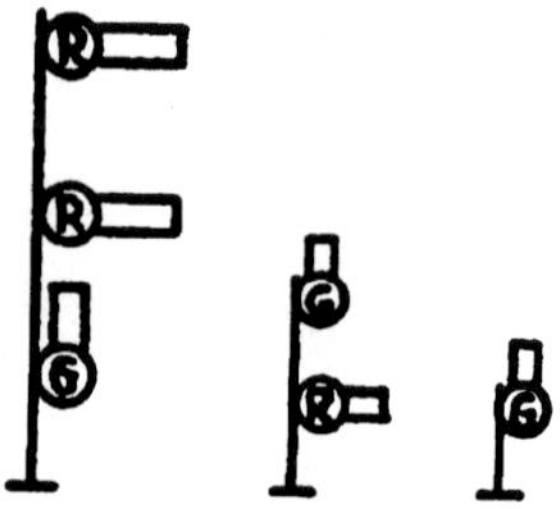

A B C

Indication—Proceed; slow speed within interlocking limits.

Name: Slow—clear.

RULE 288

Indication—Proceed preparing to stop at next signal; slow speed within interlocking limits.

Name: Slow—approach.

RULE 289

Designate by:

1—Letter plate, or

2—Marker light, or

3—Shape of arm, or

4—Combination of these distinguishing features.

Indication—Manual Block occupied; proceed prepared to stop short of train or obstruction, but not exceeding 15 miles per hour.

Name: Permissive.

RULE 290

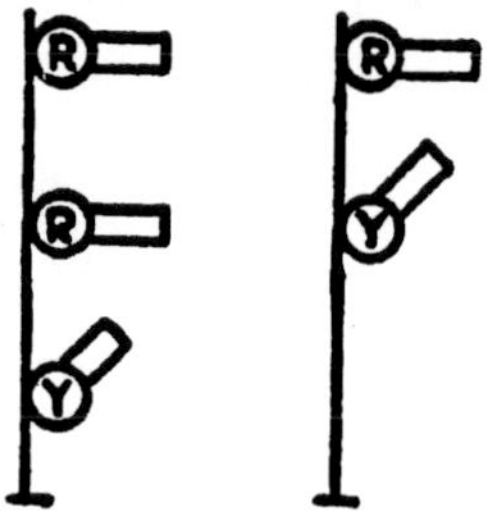

A B

C D

Indication—Proceed at restricted speed.

Name: Restricting.

RULE 291

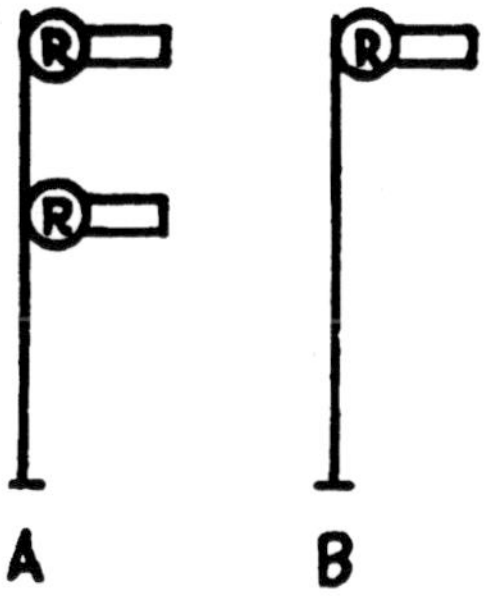

Designate by:
1—Number plate, or
2—Marker light, or
3—Pointed blade, or
4—Combination of these distinguishing features.

Indication—Stop; then proceed at restricted speed.

Name: Stop and proceed.

NOTE—Railroads desiring to avoid stopping trains may arrange accordingly.

RULE 292

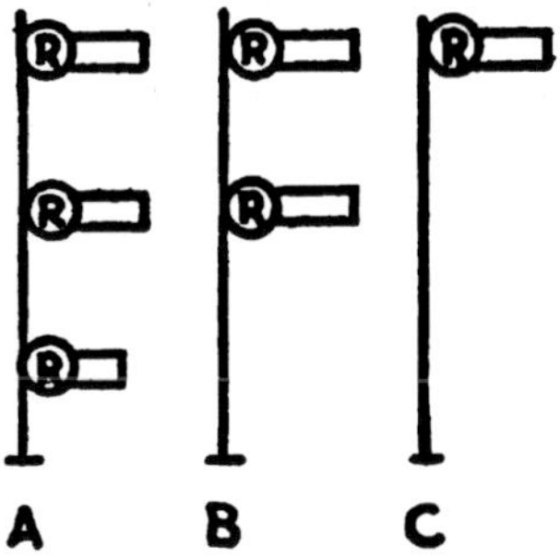

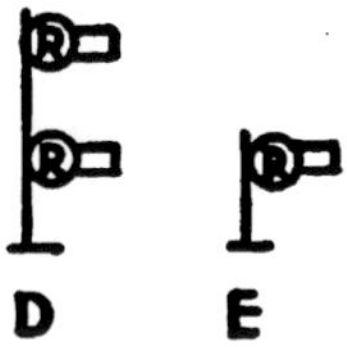

Indication—Stop.
Name: Stop.

293. A signal imperfectly displayed, or the absence of a signal at a place where a signal is usually shown, must be regarded as the most restrictive indication that can be given by that signal, except that when the day indication is plainly seen, or when sufficient lights in a position or color position light signal are displayed to determine indication of the signal, it will govern.

A signal imperfectly displayed or the absence of a signal at a place where a signal is usually shown, must be promptly reported to the ....

294. All members of engine and train crews must, when practicable, communicate to each other by its name the indication of each signal affecting the movement of their train or engine.

MANUAL BLOCK SYSTEM

RULES

305. Block signals govern the use of the blocks, but, unless otherwise provided, do not supersede the superiority of trains; nor dispense with the use or the observance of other signals whenever and wherever they may be required.

306. When a block station is open at an irregular hour, trains must be notified when possible by train order.

307. A train having entered a block on other than a proceed indication, must not accept a proceed indication at any intermediate block station which was closed when such train entered the block, without receiving Clearance Form A.

308. When Clearance Form is used information will be shown as to condition of block "clear" or "occupied." If block is occupied train will proceed prepared to stop short of train or obstruction, but not exceeding 15 miles per hour.

308. (Optional). When Clearance Form A is used information will be shown as to condition of block "clear" or "occupied." If block is occupied train will proceed prepared to stop short of train ahead.

OPERATORS

311. Signals must be kept in the position displaying the most restrictive indication, except when displayed for an immediate movement.

312. Appliances must be operated carefully and only by those charged with that duty. If any irregularity affecting their operation is detected the signals must be displayed to give their most restrictive indication until repairs are made.

313. Operators must observe, as far as practicable, whether the indications of the signals correspond with the positions of the levers.

314. Operators must not make nor permit any unauthorized repairs, alterations or additions to the apparatus.

Any defects in the appliances must be promptly reported to the . . . .

315. A block record must be kept at each block station.

317. On single track, to admit a train to a block, operator must examine block record and if block in advance is clear of opposing trains and preceding passenger trains, request operator at block station in advance to "Block for No. . . . ." If it is proper for the train to be admitted, the operator in advance will reply "OK, blocking for No. . . . ." Operator requesting block will then display the proper signal indication. Operators must promptly enter on block record the time block is given or received.

A passenger train must not be admitted to an occupied block, nor any train admitted to a block occupied by an opposing train or a passenger train, except under flag protection.

A train other than a passenger train may be permitted to follow a train other than a passenger train into a block under Permissive indication, train order or Clearance Form A.

317. (Optional). On single track, to admit a train to a block, operator must examine block record and if block in advance is clear of opposing trains and preceding passenger trains, request operator at block station in advance to "Block for No. ...." If it is proper for the train to be admitted, the operator in advance will reply "OK, blocking for No. ...." Operator requesting block will then display the proper signal indication. Operators must promptly enter on block record the time block is given or received.

A train must not be admitted to a block which is occupied by an opposing train or by a passenger train except as provided in Rule 333 or by train order. A train may be permitted to follow a train other than a passenger train into a block under Permissive indication or Clearance Form A.

318. On two or more tracks, to admit a train to a block, the operator must examine the block record and display proper signal indication.

A passenger train must not be admitted to an occupied block, nor any train admitted to a block occupied by an opposing train or a passenger train, except under flag protection.

A train other than a passenger train may be permitted to follow a train other than a passenger train into a block under Permissive indication, train order or Clearance Form A.

318. (Optional) On two or more tracks, to admit a train to a block, the operator must examine the block record and display proper signal indication.

A train must not be admitted to a block which

NOTE TO RULE 317 (OPTIONAL).—Railroads that do not require blocking opposing trains should omit this rule.

is occupied by a passenger train except as provided in Rule 333 or by train order. A train may be permitted to follow a train other than a passenger train into a block under Permissive indication or Clearance Form A.

319. When a train enters a block, the operator must give train number and time to the next block station in advance. When two or more tracks are used in the same direction, operators must also specify the track. When the rear of the train has passed ...... feet beyond the Block Signal, he must give the record of the train to the next block station in the rear.

This information must be entered on the block records.

320. Unless otherwise provided, operators must not ask for the block until they have received a report of the train from the next block station in the rear.

321. Operators must, as far as practicable, observe all passing trains and note whether they are complete and in order. Should a train without markers pass a block station, the operator must notify the operator at the next block station in each direction, and must not report that train clear of the block, nor clear the signal for the block in the rear, until he has ascertained that the train is complete.

322. Should a train pass a block station with any indication of conditions endangering the train, or a train on another track, the operator must immediately notify the operator at the next block station in advance, and each must display Stop-indications to all trains that may be affected, and must not permit

---

NOTE TO RULE 318 (OPTIONAL).—Railroads that do not use Rule 317 (Optional) should omit "On two or more tracks."

any train to proceed until it is known that its track is not obstructed.

325. An operator informed of any obstruction in a block must immediately notify the operator at the other end of the block, and each must display stop-indications to all trains that may be affected until it is known that the track is unobstructed.

326. When a train or engine clears the main track, the operator must know it is clear of the block before giving block for another movement or displaying a clear indication for that block.

The operator must obtain control of the block before permitting a train or engine to re-enter the block.

327. To permit a train or engine to enter a block or foul the main track, or to cross from one main track to another, the operator must examine the block record and if all the blocks affected are clear of approaching trains he must provide or arrange for block protection before permission is given, and until movement is complete and block clear, trains will not be admitted to the block except as prescribed by Rule 289 or Clearance Form A.

All crossover movements must be entered on the block record.

328. When coupled trains are separated, as prescribed by Rule 364, the operator must regard each portion as an independent train.

329. When necessary to stop a train for which other than a Stop-indication has been displayed and accepted, the operator must give hand signals in addition to displaying the stop-indication.

333. When, from any cause, an operator is unable to

communicate with the next block station in advance, he must stop all trains approaching in that direction. Should no cause for detaining a train be known, it may then be permitted to proceed with Clearance Form A, provided .... minutes have elapsed since the passage of the last preceding train.

334. Hand signals must not be used when the proper indication can be displayed by the block signals, except as prescribed by Rule 329 or 343. When hand signals are necessary they must be given from such a place and in such a way that there can be no misunderstanding on the part of enginemen or trainmen as to the signals, or as to the train or engine for which they are intended.

335. Block signals for a track apply only to trains moving with the current of traffic on that track. Operators will use .... for blocking trains moving against the current of traffic.

336. Operators are responsible for the care of the block station, lamps, and supplies, and, unless otherwise provided, of the signal apparatus.

337. Lights in block stations must be so placed that they cannot be seen from approaching trains.

339. If a Stop-indication is disregarded, the fact must be reported to the next block station in advance and then to ....

340. To open a block station, the operator must notify the next block station in each direction and record the trains that are in the extended block.

When trains, which were in the extended block when the block station was opened and which had passed his block station before it was opened, clear the block in advance he must repeat the record to the block station in the rear.

341. A block station must not be closed except upon authority of ....

342. Unless otherwise provided, a block station must not be closed until the block in each direction is clear of all trains.

To close a block station, the operator must notify the next block station in each direction and when he receives the acknowledgment enter it on the block record, with the time it is received from each block station.

The block signals must then be secured in the clear position, all lights extinguished, and the block wires and, when necessary, circuits arranged to work through the closed block station.

343. When a block station is open at an irregular hour, operators must use hand signals, in addition to block signals, to give the required indications until all trains have passed which have not been notified by train order that the block station is open.

344. Operators must not permit unauthorized persons to enter the block station.

ENGINE AND TRAIN CREWS

361. Block signals for a track apply only to trains moving with the current of traffic on that track.

.... will be used for blocking trains moving against the current of traffic.

362. Trains must not pass a Stop-indication without receiving Clearance Form A.

363. Train must not proceed on hand signals as against block signals.

364. Unless otherwise directed, when two or more trains have been coupled and so move past any block sta-

tion, they must be separated only at a block station and the operator notified.

365. When a train or engine clears the main track, the conductor or engineman must report "Clear" to the operator unless the switch involved is controlled by the operator. A train or engine must not enter a block or foul the main track, or cross from one main track to another, without permission of the operator.

A train or engine having passed beyond the limits of a block must not back into that block without permission of the operator.

370. When there is an obstruction between block stations, notice must be given to the nearest operator.

371. When a train is stopped by a home or block signal the conductor or engineman must, when practicable, immediately ascertain the cause.

372. Conductors must report to .... any unusual detention at block stations.

373. A block station must not be considered as closed, except as provided for by time-table or ....

AUTOMATIC BLOCK SYSTEM RULES

505. Block signals, cab signals, or both, govern the use of the blocks, but, unless otherwise provided, do not supersede the superiority of trains; nor dispense with the use or the observance of other signals whenever and wherever they may be required.

ENGINE AND TRAIN CREWS

508. Unless otherwise provided, block signals for a track apply only to trains or engines moving with the current of traffic on that track.

509. When a train is stopped by a Stop-indication it must stay until authorized to proceed, and will then proceed at restricted speed, or in case of failure of means of communication it may proceed when preceded by a flagman to the next signal displaying a proceed indication.

510. When a train is stopped by a block signal which is evidently out of order, unless otherwise provided, the fact must be reported to ....

511. Both switches of a crossover must be open before a train or engine starts to make a crossover movement, and the movement must be completed before either switch is restored to normal position.

512. Where switch indicators are used, the indications displayed do not relieve enginemen and trainmen from protecting their train as required by the rules.

513. Before a train or engine enters on or fouls a main track, or crosses from one main track to another, at bolt-locked switches, after operating the bolt-lock, the trainmen must wait .... minutes before operating the switch; at non-bolt-locked switches trainmen will operate the switch and wait .... minutes at the switch before making engine or train movement. This will not relieve employes from the duty of promptly and properly protecting the movement.

513 (a). When a train or engine has passed a signal permitting it to proceed and is delayed in the block, it must proceed at restricted speed to the next signal.

513 (b). Unless otherwise protected, entrance to a main track through a spring switch must be made under the provisions of Rule 513, operating the switch by hand.

514. A train or engine entering a block between signals must be protected as required by the rules and must proceed at restricted speed to the next signal in advance.

515. A train or engine having passed beyond the limits of a block must not back into that block except under protection as prescribed by Rule 99 or . . . .
. . . .

516. Should cab signal and fixed signal indications conflict, the more restrictive indication will govern.

517. Cab signal indications do not supersede fixed signal indications, except when cab signal changes to a more restrictive or a more favorable indication after passing a fixed signal.

518. If after passing a fixed signal, the cab signal indication changes from restricted speed to a more favorable indication, speed must not be increased until train has run its length.

CHAPTER IV

INTERLOCKING RULES

605. Interlocking signals govern the use of the routes of an interlocking, and as to movements within Home Signal limits, their indications supersede the superiority of trains, but do not dispense with the use or the observance of other signals whenever and wherever they may be required.

Rules 99 and D-152 do not apply within Home Signal limits.

606. **EMERGENCY SIGNALS**

(Whistle or Horn)

Note.—The signals prescribed are illustrated by "o" for the short sounds; "———" for the longer sounds.

SOUND	INDICATION
(a) ———	All trains and engines within interlocking limits stop immediately.
(b) o o	Resume normal movement after receiving the proper signal or permission from the operator.
(c) o o o	Whistle or horn test.
(d) o o o o	Call signal maintainer or repairman.

OPERATORS

611. Signals must be kept in the position displaying the most restrictive indication, except when displayed for an immediate movement.

612. Appliances must be operated carefully and only by those charged with that duty. If any irregularity

affecting their operation is detected the signals must be displayed to give their most restrictive indication until repairs are made.

613. When the route is set the signals must be operated sufficiently in advance of approaching trains to avoid delay.

615. When necessary to change any route for which the signals have been cleared for an approaching train or engine, switches, movable frogs or derails must not be changed or signals cleared for any conflicting route until the train or engine for which the signals were first cleared has stopped.

616. The lever operating a switch, derail, movable frog, detector bar or lock must not be moved when any portion of a train or engine is standing on or closely approaching the switch, derail or movable frog.

617. Operating levers must be blocked or marked and should not be used when a track, switch or signal is undergoing repairs or when a track is obstructed.

619. If the force whose duty it is to keep switches clear when snow or sand is drifting, is not on hand when required, the fact must be reported to the . . . .

620. If a signal fails to work properly its operation must be discontinued and until repaired the signal secured so as to display its most restrictive indication.

621. Operators must observe, as far as practicable, whether the indications of the signals correspond with the positions of the levers.

622. Operator must not make nor permit any unauthorized repairs, alterations or additions to the interlocking.

Any defects in the interlocking must be promptly reported to the . . . .

523. If there is a derailment or if a switch, movable point frog or derail is run through, or if any damage occurs to the track or interlocking, the signals must be restored so as to display their most restrictive indication, and no movement permitted until all parts of the interlocking and track liable to consequent damage have been examined and are known to be in safe condition.

524. Whe necessary to disconnect a switch, movable point frog, derail, facing point lock, detector bar or electric locking circuits, all switches, movable point frogs and derails affected must be securely spiked or fastened in the required position and the levers blocked or marked in such a manner that they cannot be operated, before any train or engine is permitted to pass over them.

525. When switches, movable point frogs, derails or signals are undergoing repairs, Stop-indication must be displayed for any movement which may be affected by such repairs, until it has been ascertained from the repairman that the switches, movable point frogs and derails are properly lined for such movement.

526. Operators must, as far as practicable, observe all passing trains and note whether they are complete and in order; should there be any indication of conditions endangering the train, or any other train, the operator must take such measures for the protection of trains as may be practicable.

528. Hand signals must not be used when the proper indication can be displayed by the interlocking signals.

529. If necessary to authorize a train or engine to pass an interlocking signal indicating stop, hand signal

or permission may be given by the operator. Such occurrence must be reported to the ....

630. Operators are responsible for the care of the interlocking station, lamps and supplies.

631. Lights in interlocking stations must be so placed that they cannot be seen from approaching trains.

633. If a train or engine overruns a Stop-indication, the fact must be reported to ....

634. Operators must not permit unauthorized persons to enter the interlocking station.

635. During a stated period an interlocking station may be closed upon authority of .... When so closed, switches and switch levers must be secured for routes that do not conflict and signal levers placed in position so that signals will display the proper indication.

The interlocking station must be securely locked.

636. At interlocking stations, where there is a train order or block signal operated by the same operator, such signals must not be changed to display a proceed indication for a train until after the interlocking signal has been changed to permit the train to proceed.

ENGINE AND TRAIN CREWS

661. If a signal indication, permitting a train or engine to proceed, after being accepted, is changed to a Stop-indication before it is reached, the stop must be made at once. Such occurrence must be reported to ....

663. Trains or engines must not pass an interlocking signal indicating stop until a member of the train

or engine crew is fully informed of the situation. Movement may then be made on hand signal or permission of the operator, at restricted speed.

667. Sand must not be used nor water allowed to run over movable parts of an interlocking or spring switch.

669. Trains or engines stopped by the operator in making a movement through an interlocking, must not move in either direction until they have received the proper signal from him.

670. A reverse movement within the limits of an interlocking or a forward movement after making a reverse movement, must not be made without the proper interlocking signal indication or permission from the operator.

670 (a). When a train or engine having accepted a proceed indication of an interlocking signal, stops less than 30 feet in advance of such signal, it must not again proceed without permission from the operator.

671. While an interlocking station is closed, should a signal for an open route indicate "Stop," movements through the interlocking must be preceded by a flagman. Before proceeding, the engineman and trainmen must know the route is properly lined.

The facts must be reported to the .... from the first available point of communication.

672. When a train or engine is stopped by a Stop-indication of an automatic interlocking signal and no immediate conflicting movement is evident, a member of the crew must operate the time release. If signal does not change its indication at expiration of time release interval, train or engine may then proceed on hand signal from a member of the

crew if there is no train or engine on conflicting route and signals on conflicting route indicate Stop.

If a train or engine is on conflicting route, hand proceed signal must not be given until such movement is stopped, and if signals on conflicting route do not indicate Stop, proper flag protection must be provided.

Part II

Chapter V

APPLICATION OF THE STANDARD CODE ON INDIVIDUAL ROADS

Special attention is called to the Code provision: "Where proper safeguards are provided, railroads may modify or add to these rules to suit their requirements."

Many blank spaces will be found in these rules; and the student is referred to his own rule book. There is an increasing tendency on some carriers to reshuffle the rules themselves, even to the point where the numbers do not agree with the Standard Code.

The physical characteristics of a road, its peculiar method of operation, and the ideas of the Rules Committee on each road are all factors. However, the rules of each road dovetail together in a manner to provide "proper safeguards," even though the individual rules may differ from book to book.

This section will be devoted to showing how this is so. In some cases radical departures from the Standard Code will be found. The beginner is cautioned to make a careful study of the rules of his individual road. They may differ from anything presented herein. Obviously, it is impossible to include an analytical study of the rules of each individual carrier in a book of this kind, to show how they fit together into a pattern of safety.

By study of the illustrative examples the student will broaden his understanding of the principles involved. When he grasps the reasoning behind the rules he can apply the principles to the rules of any road. He should then be able to solve unassisted any rules problems he is likely to encounter.

The importance of the definitions cannot be overstressed. Unless the student understands precisely what is meant by the word "train," for instance, the meaning of a rule may be entirely missed. One of the most widely discussed rules problems of which I have any recollection arose out of a misunderstanding as to the definition of the word "train." Some claimed that a "section" of a schedule was not a "train," and it took an A. of A. R. decision to settle the question for good: That anything which conformed to the definition of a "train" was really a "train," regardless of the authority under which it moved.

Time is well spent in memorizing the definitions. Then, if any rules is not entirely clear, stop and substitute the definition for the term used, and the meaning of the rule is clarified.

One thought must be kept uppermost in mind: A railroad cannot be operated safely unless each employe is conversant with and obeys the rules. Not even ninety-nine percent of perfection will be sufficient. They provide a specific course of action under all circumstances, which, if obeyed, produces coördination. If a man is ignorant, or neglectful, and does something in violation of the rules, others then become confused and hazard arises. He may cause the death of others who know their rules and their rights under them.

If any employe believes that certain rules are unimportant or unreasonable, he must be made to understand why he is wrong. Otherwise, some form of discipline must compel his obedience to the rules to protect the lives of others as well as his own life.

When a question arises as to the reason for a rule, or its exact meaning or application, the proper authority should be consulted for an explanation. In the original Standard Code, there was a clause in the General Notice to the effect that: "If rules cannot or ought not to be enforced, they should not exist." This axiom is as pertinent today as it was in 1899.

Rules may be found on some roads today "which do not permit a man to do a job of railroading"; however, these are the exception, not the rule. Should you conclude that the rules of your road prohibit your doing work safely and efficiently, the thing to do is to try to get the rules changed. Do not violate them. Safety is of the first importance; and obedience to the rules is essential to safety.

Habit is of great importance. It is just as easy to do things right as to do them haphazardly. Without strict observance of regulations, an army would be a mob. The same requirement applies to a railroad. Each employe must know what the other fellow will do under a given set of circumstances. The best manner in which this understanding can be produced is a good set of rules, rigidly enforced.

Courtesy, which springs primarily from one's consideration for the feelings of others, brings its own reward in personal satisfaction. But it also makes and keeps friends and minimizes friction. It enhances an employe in the eyes of his associates.

A man who loses his temper is more or less insane for the time being and apt to make mistakes. Men in the operating department of a railroad must keep cool and think before acting.

GENERAL NOTICE

These have been added to by almost every carrier, each of which believes said additions have clarified the requirements on their road. Here are a few:

Southern Pacific—One of the best recommendations for promotion is that an employe has so performed his duties as to win the good will and friendship of patrons for the railroad and for himself.

Suggestions from employes intended to promote safety, economy, or service, are solicited and will receive consideration.

The public judges a railroad by the appearance and conduct of its employes, quality of service, and condition of the property. Courteous, considerate treatment of patrons is of the first importance in retaining and increasing our volume of business, and also governs the extent of opportunity for employment in the railroad's service.

Santa Fe—Employes, in accepting employment, assumes its risks.

Chicago and North Western—In case of doubt adopt the safe course. Speed must always be sacrificed for safety.

The following rule from the New York Central, illustrates very well how each road writes those rules essential to the operation of its property, and why the rules differ from road to road:

"Avoid stepping, sitting, walking upon or brushing against the third rail. While the energized rail is protected by sheathing, there is always chance of shocks because of the presence of water, brake-shoe dust, derangement or imperfections of the sheathing. Employes must caution passengers and the public accordingly."

Obviously, the rule would be of no benefit on roads having no electric locomotives.

SPECIAL INSTRUCTIONS

The provision that "Special instructions may be issued by proper authority" covers a great many practices in railroad operation.

The Book of Rules is the foundation. Special instructions in the time-table supersede those of the Book of Rules where they conflict. Then, the rules of most roads provide for bulletins, whose instructions supersede those in either the time-table or rule book.

Bulletin instructions continue during the life of the current time-table, at which time, theoretically, their contents are transferred to the new time-table.

Some roads provide for Special Notices, also, which continue in effect indefinitely, regardless of any time-table change. Since the method used differs from road to road, it is necessary to examine the rule book carefully in order to understand precisely the method used on your road.

The system used on the Pennsylvania Railroad, for instance, is a departure from generally accepted practice; their instructions being designated as GENERAL ORDERS. Here are the rules governing their use:

75. General Orders will be issued by authority and over the signature of the superintendent. General Orders must be numbered consecutively, the number being prefixed by the number of the time-table. Each General Order will contain a subheading indicating in what general order zone or zones the order applies.

75a. New rules, annulments and changes in rules approved for system practice, will be made effective by General Orders on sticker coupons which the employes must insert in their copy of Operating, Signal and Interlocking Rules.

75b. Conductors and enginemen, when reporting for duty, must examine the bulletin boards to see that they are familiar with all effective general orders pertaining to any portion of the general order zones on which they are qualified. They must insert a sticker copy of all general orders in their time-tables, after which they, as well as trainmen and firemen, must sign the employes' register in the presence of a designated employe who must personally witness the signature of each employe.

Train and engine crews after having been off duty one hour or more, must register.

Conductors and enginemen reporting for duty at a place where there is no designated employe on duty, or where no general orders are posted, or where the general orders posted do not cover the territory over which they are assigned to run, must report to the operator and receive instructions by train order covering the general orders not in their possession, but which are effective in that territory.

Before starting on a trip or tour of duty, the conductor and engineman must know that the other is qualified and that he has inserted in his time-table the necessary general orders for such trip or tour of duty; also the conductor must know with respect to the trainmen, and the engineman, with respect to the fireman, that they have seen and are familiar with such general orders. Conductors and enginemen ordered to run over any portion of a division or foreign railroad over which they are not qualified must inform the superintendent of the division or foreign railroad that they are not qualified.

The general order zones on which conductors and enginemen are qualified must be shown in the space provided therefor on their time-tables.

Each qualified conductor and engineman, whether or not serving in that capacity, must have with him while on duty the time-tables of the divisions on which he is qualified in part or whole, showing the general order zones in which he is qualified and containing the necessary general orders pertaining thereto.

Train orders, of course, must be understood as superseding any instructions with which they conflict. For example, a train order may prescribe a speed of four miles per hour over track which, according to the time-table, has a permissible speed of sixty miles per hour. Time-table may show a meeting point between Nos. 1 and 2 at B and a train order may fix the meeting point at C instead and may even reverse the time-table superiority of trains or annul them.

The Southern Pacific rule is somewhat different and reads as follows:

> Bulletin boards or books will be provided at stations designated by time-table. Time-table bulletins will contain only information or instructions relating to the rules or movement of trains, supersede special instructions in the time-table or any rule or regulation with which they conflict, and expire with the current time-table.
>
> Conductors, engineers, yardmen, and others concerned in the movement of trains must review time-table bulletins and instructions before commencing each day's work.
>
> Time-table bulletins will be issued by authority and over the signature of the superintendent, and will be numbered consecutively during the effectiveness of each time-table, and, in addition to the bulletin number, will bear the number of current time-table.

Where bulletins are posted on bulletin boards or in bulletin books, there is usually some plan to supply conductors with copies of certain bulletins to be placed in books and kept where they may be easily referred to while on the road. This enables conductors of work trains or others who may not leave initial stations daily to keep posted.

The definitions are the keys to the rules. A great many of them will be found to be identical with the Standard Code. However, the individual rule book must be studied carefully and the definitions memorized if one is to be able to understand, or apply the rules intelligently to a problem.

As equipment and operating practices are improved, the definitions must be altered so that no confusion will arise as to the precise meaning of the terms used in the rules. For instance, with the coming of diesel-electric power, the old definition of an engine, which read: "A machine propelled by any form of energy and used in train or yard service," did not entirely convey the desired meaning. Diesel engines are usually composed of several units operated from a single control.

So the definition was changed to read: "A unit propelled by any form of energy, or a combination of such units operated from a single control, used in train or yard service."

The word "unit" is still applicable to the steam engine, making the definition cover all types of motive power. Had the phrase, "or a combination of such units operated from a single control," been left out of the definition, then confusion might have arisen as to whether a four-unit diesel was actually one engine, or four engines.

A few years ago, the term "remote control" was unknown in railroading. With the coming of Centralized Traffic Control, where signals and switches are remotely controlled, a definition had to be provided to give the term an exact meaning: "A term applied to a system of operating outlying switches or signal appliances from a designated point." Incidentally, the Standard Code does not contain such a definition.

CTC also brought with it another term which had to be defined: "DUAL CONTROL SWITCH—A power-operated switch which is also equipped for hand operation."

In addition to such changes as these, the method used in train operation on various carriers have caused them to alter their definitions with regard to speed. To illustrate, the definitions from various roads are given below:

PENNSYLVANIA

MEDIUM SPEED—Not exceeding one-half the speed authorized for passenger trains but not exceeding 30 miles per hour.

REDUCED SPEED—Prepared to stop short of train or obstruction.

SLOW SPEED—Not exceeding 15 miles per hour.

RESTRICTED SPEED—Not exceeding 15 miles per hour, prepared to stop short of train, obstruction or switch not properly lined and to look out for broken rail.

B. & O.

NORMAL SPEED—The maximum speed permitted by time-tables for main track movements.

MEDIUM SPEED—One-half the normal speed, not to exceed thirty (30) miles per hour.

SLOW SPEED—One-quarter of the normal speed not to exceed fifteen (15) miles per hour.

RESTRICTED SPEED—Proceed, prepared to stop short of train, obstruction, or anything that may require the speed of a train to be reduced.

UNIFORM CODE (M. P. & OTHERS)

MEDIUM SPEED—A speed not exceeding 30 miles per hour.

RESTRICTED SPEED—Proceed prepared to stop short of train, engine, obstruction or switch not properly lined.

LOW SPEED—A speed that will permit stopping short of train, engine, obstruction or switch not properly lined and looking out for broken rail, but not exceeding 15 miles per hour.

S. P.

WITH CAUTION—To run at reduced speed, according to conditions, prepared to stop short of a train, engine, car, misplaced switch, or other obstruction, or before reaching a stop signal. Where circumstances require, train must be preceded by a flagman.

These examples show clearly how the rules of one road differ from another; and how they may also differ from the Standard Code. Therefore, the student MUST consult his own rule book to determine the precise wording of the rules on the road for which he works. It may have different methods of operation; therefore, different definitions and rules. A book of this kind cannot take up in detail the rules of each individual carrier. The best that can be done is set forth in the principles involved to enable the student to better interpret the rules on the road where he is employed.

On roads where signal control supersedes the superiority of trains, or the time-table superiority of trains, speed definitions must be written to give the exact meaning to the rules which, in turn, give meaning to the particular system of moving train involved.

RULES 1, 2 AND 3

The Standard Code leaves each road to determine those employes subject to the provisions of Rule 2. The period of time elapsing between filing of certificates also varies from road to road. Employes usually listed, in addition to trainmen and enginemen are: Hostlers, train order and signal operators, signal maintainers, linemen, track and bridge foremen, yardmen, yardmasters, flagmen, train dispatchers and minor officials.

These vary from road to road, the outstanding departure from usually accepted practice being on the Pennsylvania. Their Rule 2 reads: "Conductors, enginemen and other designated employes must use reliable watches." The reason for lowering of the usual standards will be found in their system of handling trains. The bulk of train movements are made by signal control; and block signals protect movements made by train order. Time orders are held to an absolute minimum, the "meet" order being generally used.

Here is Rule 3 from the Uniform Code effective on the Missouri Pacific and other roads, May 1, 1950:

> 3. TIME COMPARISON—Standard Clocks must bear sign "Standard Clock" and their locations will be shown in special instructions. Where standard clocks are available, watches of train dispatchers, conductors, engineers, hostlers and engine foremen, must be compared with a standard clock before commencing each day's work or trip. The time when watches are compared must be registered on a prescribed form.
>
> Conductors, engineers, hostlers, engine foremen and train-order operators, whose duties preclude access to a standard clock, must obtain time from train dispatcher, or compare and regulate their watches

daily with those of conductors and engineers who have standard time and have registered as provided.

When practicable, conductors or engine foremen, and engineers will also compare time with each other and with trainmen, or yardmen, and firemen before commencing each day's work or trip.

RULE 4

There is at least one road which has eliminated the definition of "subdivision"; and this word does not appear in Rule 4. On other roads, the definition of "division" is retained but the word does not appear in Rule 4. The student is cautioned to examine his book of rules carefully.

Where subdivisions actually exist, the elimination of "division" from Rule 4 has no effect; however, the elimination of the word "subdivision," where they exist, has a profound effect upon Rule 4, as will be shown.

In the past, some roads employed the term "district" instead of "subdivision"; and it may be that this practice still exists on these roads. The words have an identical meaning: "A portion of a division designated by timetable."

It will be noted that the Standard Code Rule 4 gives but six points in which two schedules must correspond, while some of the newer rule books give a seventh point. These points are: number, class, day of leaving, direction, route, and initial and terminal stations. To make the rule as clear as possible, the seven requirements will be referred to in following matter. On roads where "route" is not one of the requirements, as in the Code rule, "direction" is interpreted as also covering "route."

The circumstances attending the changes of timetables throughout the country are at such variance it would be difficult to anticipate all conditions and cover each specifically by illustration. However, sufficient material is set forth herein concerning the principles of Rule 4 as to enable anyone to learn to determine positively just what trains can, or cannot, run under the new time-table. Once this is known, the authority of other trains concerned to move ahead of, or against schedules which are affected by the change should be easily determined.

The best way to arrive at a decision, ordinarily, as to the existence or application of the time of a schedule of the new time-table is to reason as follows:

Do the schedules agree in the seven particulars?

NUMBER
CLASS
DAY OF LEAVING
DIRECTION
ROUTE
INITIAL STATION
TERMINAL STATION

Has today's schedule been fulfilled already, or so disposed of otherwise that it cannot be assumed? Is the schedule authorized—that is, due, or overdue, not to exceed twelve hours, on the road at the time of change? If not, then the schedule does not become effective until its leaving time at its initial station after the new time-table has taken effect.

If Saturday's schedule is involved, do not let it become confused with Friday's or Sunday's schedule. Stick to Saturday.

The word "direction" means northward, southward, eastward, or westward.

"Route" means that the trains must follow the same piece of main track; but it does not follow that the simple addition of a new station, or siding, or the dropping of same from the schedule would alter its "route." If in doubt, application should be made to the proper authority for an explanation.

A schedule takes its date from its initial station on each division, if not subdivided; and from its initial station on each subdivision, or district, where such exist.

It is not necessary that a schedule of the old and new time-tables correspond in day of arriving.

If a schedule on the old time-table is due to leave its initial station at the exact minute the new time-table becomes effective, there is no schedule existing under the old time-table at the moment of change. The ruling of the A. of A. R. as to this reads:

> On Time-table No. 1 Train 88 is due to leave "A" at 1:00 o'clock a.m. Time-Table No. 2 becomes effective at 1:00 a.m. June 10th, and it shows No. 88 due to leave "A" at 12:01 a.m. Train 88 on Time-Table No. 2 corresponds in number, class, date of leaving, direction, initial and terminal stations with the same train on Time-Table No. 1. The question is: Is Train 88 authorized to run on the date that the new time-table becomes effective?

> While schedule 88 on Time-Table 2 corresponds in number, class, day of leaving, direction, initial and terminal stations with schedule 88 on Time-Table 1, the schedule was not in effect at the time of change, so that no train could be authorized to assume it. Therefore No. 88 was not authorized to run on the date the new time-table became effective, and cannot start from its initial station until 12:01 a.m. the following day.

If a schedule of the new time-table is due at its initial station at the exact minute the new time-table takes effect, it comes into existence simultaneously with the new time-table; therefore, it is in effect at the time of change.

Remember that a schedule of the old time-table must be in effect at the time of change. For instance, if the old time showed No. 1 due to leave its initial station at 1:00 p.m., and the new time-table became effective at 12:01 p.m. Sunday, scheduling No. 1 out of its initial station at 11:00 a.m., there could be no No. 1 of Sunday. The first No. 1 that could exist under the new time-table would be No. 1 of Monday, the first leaving time after the new time-table became effective.

Now, if No. 1 on the old time-table had been scheduled to leave at 11:59 a.m., there would be a train authorized at the time of change, which would change the situation. No. 1 could then assume the schedule of the corresponding number in the new time-table, leaving at 12:01 p.m., one hour and one minute late, provided the schedules agreed as required.

Again, there are times when the schedules agree in the seven requirements and a train will lose both right and schedule because, when it assumes the schedule in the new time-table, it finds it is more than 12 hours late.

These variations, or combinations, are best illustrated by example; and the student is referred to the Question and Answer section on Rule 4, where almost every conceivable combination is presented for study.

The majority of roads have added to Rule 4 the conditions under which the change of time-table will be made. These vary from road to road. Here is a rule from the Santa Fe Book of Rules:

4 (a). Notice of the issuance of a new time-table will be posted in bulletin books. Train dispatcher will notify yardmasters, conductors and enginemen by train order of the effective date of a new time-table.

This order must be in effect for twenty-four hours in advance and seventy-two hours after the new time-table is effective.

There are two general methods: one on roads which simply deliver a copy of a "19" order, or "train order" where only one form is used, which require no signature, reading: "Time-table No. ...... effective ...... A. M. .... (date) .....

The other is to issue a "31" order reading: "I have time-table No. ...... effective ......," which is handled in the usual manner for "31" orders.

RULE 5

Most revised rule books conform closely to Code Rule 5. Some roads adhere to the old method. They call attention to where there are one or more trains to meet or pass a train between two times, or more than one train to meet or pass a train at any station by using the heavy dash under the figures, thus:

5.45	or	7.10
▬		
		7.30
		▬

Other roads place the number of the schedules to be met or passed adjacent to the full-faced, or heavier, type at all meeting and passing points, whether one or more trains are to be met or passed. This saves time in cross-checking the time-table to find what schedules are to be met or passed when they find a time printed in full-faced type.

The second paragraph of Rule 5, as amended November 15, 1938, has been considerably shortened in form without disturbing the principles involved. Time still applies to the switch where an inferior train enters the siding. Where there is no siding, it applies at the station; which, in actual operation, would be the place from which fixed signals are operated, if any; or the place where traffic is received or discharged. The shortening of the "station" definition makes the shortening of this paragraph possible.

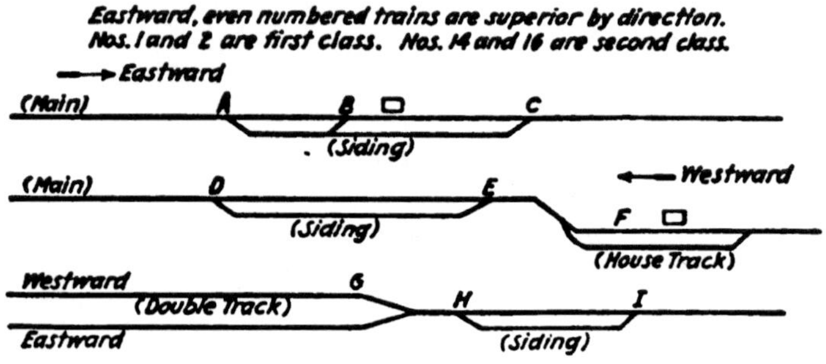

APPLICATION OF TIME

Time applies for No. 2 at C, except that no train which stops for passengers may leave the station building until due to leave the station. If No. 2 does not stop for passengers, it may go to C at any time and wait there.

Time applies for No. 1 at A. No. 1 must not attempt to go to B and pull into the siding, except under protection of flag.

If No. 16 is to let No. 2 pass, it must pull into the siding at A. If it holds the main track, it must be protected as per Rule 99. No. 2 should approach A expecting to find No. 16 pulling into clear; however, No. 16 should exercise care by having out ample flag protection.

Being of the same class, No. 16 is not required to wait at a schedule passing point beyond the time when No. 2 becomes due, unless No. 2 is seen approaching or other information is received indicating they will arrive shortly.

The second illustration shows the problems which arise when the siding is entirely to one side of the station. A "middle order" placed at this point would afford little if any protection to an eastward train, except that the

EASTWARD	—			SUBDIVISION NO. 1					—			WESTWARD
Third Class		Second Class		First Class			First Class		Second Class		Third Class	
28	26	16	14	4	2		1	3	13	15	25	27
Leave Mond. Wedn. and Frid.	Leave Daily	Leave Daily Except Sunday	Leave Daily	Leave Daily	Leave Daily	Stations	Arrive Daily	Arrive Daily	Arrive Daily	Arrive Daily Except Sunday	Arrive Daily	Arrive Tues. Thurs. and Saty.
	10.08 AM					Z		10.05 AM			9.55 AM	
		9.00 AM		9.00 AM 10.40 AM		P	10.30 AM	9.00 AM		9.00 AM		
9.50 AM 10.05 AM		10.00 AM			10.00 AM	K	10.00 AM					
	12.30 PM 1.30 PM					H			12.30 PM 11.30 AM			
10.45 AM						E			10.55 AM 10.45 AM			
2.00 PM						A						2.00 PM
Arrive Mond. Wedn. and Frid.	Arrive Daily	Arrive Daily Except Sunday	Arrive Daily	Arrive Daily	Arrive Daily		Leave Daily	Leave Daily	Leave Daily	Leave Daily Except Sunday	Leave Daily	Leave Tues. Thurs. and Saty.

Eastward Trains are Superior to Westward Trains of the Same Class.

train is required to stop at E (where an inferior train would enter the siding) if the train order signal were at "stop." An order restricting an eastward train would constitute a potential hazard, if placed at that station for delivery.

No. 2 must not pass E in advance of its leaving time, for the purpose of doing station work. Neither can the time consumed by No. 2 doing station work be used by an opposing train, for No. 2 is by siding switch where they would head in. If given an order to wait here, No. 2 would have to remain at E until the wait expired, then go to the station to do its work.

In the third diagram, eastward trains would go to I to wait for time, or meet an inferior train, which would head in at that point. If the dispatcher wishes an inferior train to move to the end of double track at G for a superior train, it is best that he give the inferior train right over the superior train to the end of double track at G, thus avoiding confusion as to the exact point of restriction.

Time-table time, train-order waits, or run-late orders all apply at the same points.

Considering the above time-table chart, it is not necessary to show the time of Nos. 27 and 28 in full-faced type at A, because schedule is to run on different days.

No. 13 should have only arriving time at E and leaving time at H shown in full-faced type, as trains are met on arriving and leaving time.

It may seem unnecessary to show a dash under the time of Nos. 1 and 2 at K, as they are only concerned in meeting each other. Opinions differ as to this; however, when running about on the time of a regular train in the same direction, crews of extra trains sometimes consider its schedule meeting and passing stations as a check upon their memories. For this reason, if for no other, it would seem advisable to show Nos. 1 and 2 with a dash under time at K; or, as prescribed by latest revised Code rule, show the time for No. 1 thus:

10:00 A.M. 2 16 28

No. 4 should be shown with a dash under its time at P, because No. 1 is met between its arriving and leaving time.

It is unnecessary to show the time of Nos. 3, 25 and 26 in full-faced type at Z because the time is not identical. This is in accordance with A. R. A. ruling of November 18, 1915.

RULE 6

Every company has found it necessary to add letters or signs to Code Rule 6, such as the following:

TO—train order office;
R—train register station;
B—bulletin station;
K—standard clock;
W—water station;
C—coal station;
O—fuel oil station;
F—fuel;
I—interlocking;
T—turn-table;
Y—wye or turning track;
P—telephone;
m—stop for meals;
¶—stop for meals;
D. T.—double track;
A. B. S.—automatic block signals;
A T. C.—automatic train control;
A.A.B.S.—absolute automatic block signals;
CTC—Centralized Traffic Control;
V—track connection with foreign road;
Z—track scales.

There are others besides the above, and each road does not always use the same symbol or letter to mean the same thing. One must consult his individual book.

It will be found that such letters as ABS, formerly indicating conventional automatic block signals, are now sometimes applied to a form of Centralized Traffic Control; the same being true of AABS. CTC is actually a trade name, although it has been adopted in the rules of many roads to designate the method of dispatching used.

FIXED SIGNALS

Before going into the various signal rules, it is well that we understand exactly what constitutes "fixed signals." They cover signals of fixed location, such as switch, train-order, interlocking, stop signs, yard-limit signs, slow signs or other means for displaying indications that affect the movement of a train or engine.

Generally speaking, "fixed signals" require no acknowledgement from the engineman by two short blasts of the engine whistle. Trains reduce speed, stop, proceed at various speeds according to the indications from the various signals.

One exception to this procedure is the train-order signal on roads where the normal position of the signal is "stop." If the engineman is required to call for the signal by four short blasts of the engine whistle before the operator may clear it, then the clear signal must be acknowledged by two short blasts of the engine whistle.

This leads to a "whistle nuisance" against which some municipalities have passed laws. As a result, the rules have been amended so that the whistling is eliminated in thickly populated areas, the engineer being authorized to accept the cleared signal without calling for it, or answering it. Operators are instructed to wait until they are certain the engineman can see the signal, in stop position, then clear it.

Whistling posts are fixed signals prescribing the use of the engine whistle; however, in obeying the instructions of such a signal the engineman acknowledges the indication but does not do so with two short blasts of the whistle. Roadway foremen must keep weeds cleared away from slow signals. Slow signals should be placed on straight track, where possible, so they will not be hidden by rocks, trees or curvature.

When section or bridge foremen know track or bridges are unsafe, they must not depend solely upon signals at the side of the track, but must stop trains and notify the crews. They may understand they are relieved from stopping trains and notifying crews only after they have advised dispatcher and obtained acknowledgment of the notice from him.

A good form of slow-board is one having two speeds indicated, that for passenger trains above, for other trains below. Since the days of streamliners, some roads use three sets of figures: one for streamliners, one for other passenger trains and a third for freight trains. The sign should be about on a level with the eye of the engineman and must be understood to restrict trains until another sign is reached, indicating other speeds, or until the end of the restriction, or end of the subdivision.

Watchmen must be instructed to care for lights and see that they do not become dim or fail. It must be kept in mind always that an unattended signal may fail and should not be depended upon implicitly to stop trains.

RULES 7, 8, 9 and 10

Rules 7, 8 and 9 are uniformly adopted by the various roads. However, Rule 10 is usually found with various additions, such as these taken from the Southern Pacific Lines rule book:

10 (G). When an unattended red flag or red lamp is displayed on or near track, train, after stopping, must be preceded, for a distance of three-fourths mile from the point where the signal is displayed, by a flagman who must carefully examine track and structures, and if signal is displayed on the track it must be replaced on the track after train has passed.

A signal so displayed will not apply to the track on which the train is running if displayed beyond the first rail of an adjoining track.

10 (H). When a yellow signal is required it will be displayed to the right of track in direction of approach, three-fourths mile from structure or track over which speed of trains must be restricted. Where two or more main tracks are affected the signal will be displayed for each track the same as if it were a single track. A green metal signal will be displayed to the right of each track at the limit of restriction, and trainmen will give signal 12 (c) after rear of train has passed it.

Trains must not exceed the speed specified by train order, or time-table bulletin, or fifteen miles per hour if no different speed is so specified, while passing over the structure or track affected, until the rear of train clears the restricted limit.

RULE 11

Rule 11 of the Standard Code, while used by the majority of roads, is varied some from road to road. Here is the Southern Pacific version:

11. When an unattended fusee is burning on, or near a track within block system or centralized traffic

control limits, train may proceed without stopping, but must run with caution, not exceeding fifteen miles per hour, for three fourths mile.

When a fusee is burning on, or near a track outside of block system or centralized traffic control limits, train must stop, and not proceed until fusee has burned out.

A fusec burning beyond the first rail of an adjoining main track will not apply to the track on which train is running.

11 (A). Fusees must not be placed in timber-lined tunnels, nor on bridges, road crossings, or signboards, not at places where damage from fire may result, and must not be placed between rails of a track when practicable to avoid it.

Some roads do not require passenger trains to extinguish fusees after stopping and before proceeding.

It must be understood that a red fusee affords protection only from the point where an engineman can stop after seeing it. One advantage in the use of fusees is that when an unattended fusee is seen burning, it indicates to the engineman that a train has passed that point less than ten minutes before.

Fusees should not be placed in signal case with wet flags, nor allowed to become damp from other causes. They deteriorate rapidly, and more than a reasonable supply should not be kept on hand.

If an employe has never lighted a fusee, he should be instructed prior to the time when an emergency arises. The end should always be held away from the person lighting it; and he should ascertain that it is burning well before throwing it on the ground.

CHAPTER VI

SIGNALS AND THEIR USE

RULES 12, 13 and 14

The 1938 revision of the Standard Code eliminated the signals for "Train parted." The lamp swung at arm's length was the hand signal, to be answered by three long blasts of the engine whistle, or vice versa. Many rule books still contain these signals; and there are some who question the wisdom of eliminating them.

The large-capacity air pumps, with which locomotives are generally equipped, will maintain enough air pressure in a long train line to forestall an emergency application of the brakes. Cases are cited where trains have broken in two and the head end continuing for a considerable distance before being stopped. This constitutes the same problem that has always existed, and for which the signals were created.

Some of the engine whistle signals found in multiple-track rule books do not conform exactly to those of the Standard Code, and must be examined carefully.

Many roads use communicating signal which a conductor must give approaching points where the train is restricted; meeting or waiting points. These signals vary with different roads, the Code signals having been switched around in order to conform to the ideas of the management.

Those giving signals should locate themselves so as to be seen plainly and makes the signals in such a manner as to be readily understood. The utmost care must be exercised by both trainmen and enginemen to avoid one engineman acting on a signal intended for another engineman.

On passenger trains, hand signals can usually be given uniformly as prescribed. Long freights present a somewhat different problem. However, a white flag should be used to give a proceed signal from the rear, not a red one which the flagman may be carrying. The same is true of lights.

Hand, lamp and steam-whistle signals speak a language. If a man does not speak correctly, nor articulate well, he is liable to be misunderstood. The same is true of signals.

The aim of all types of signaling is to convey information; and the more uniformly the rules are complied with by all, the nearer all approach to a perfect understanding. The whistle signals of some enginemen sound much like the language of some foreigner who never learned English.

If a flagman is to be recalled by four long blasts of the engine whistle, then do not blow four short blasts, which is to call for signals. And the flagman should not return until he has received a proper whistle signal. The engineer may actually be calling for signals because he has found an unattended red signal.

Some consideration must be given to the spacing of blasts of the engine whistle, particularly at points where the echo causes intermingling. Two short blasts of the engine whistle is an acknowledgement of a signal, not a signal to "release brakes—proceed."

There are legal reasons for blowing the public-crossing whistle properly. If the whistle valve is opened wide-open quickly and closed in the same manner, the penetrating quality of the whistle is increased by forty percent. It is only common sense that a crossing whistle be prolonged until the engine is actually on the crossing.

Should an engineman find it necessary to stop when there is the possibility of more than a momentary delay, he should whistle out a flag. Unless this is done, the flagman, not knowing the reason for the stop, may wait too long before starting back, thus creating a hazard.

The meeting-point whistle requires space between each blast. It can be heard reasonably well on a passenger train but this is not always true on a long freight. Men on the rear can watch for the puffs of steam in the daytime but this is not true at night.

The station whistle should be at least four seconds long, so that it will carry far and give ample warning.

If it is desired to move a train a short distance after having stopped at a meeting point, or where the train is restricted for another, no signal should be employed. There have been cases where an engineer took a signal intended to move the train only a short distance for "RIGHT AWAY!", as the English say—and left town, making a milepost out of the conductor and getting himself hit.

The only sensible thing to do is to walk down and tell the engineer what move is to be made while waiting for the train for which he is restricted.

RULE 15

The majority of roads have adopted Code Rule 15 without change. There are two schools of thought with regard to the use of torpedoes. One of these is portrayed by the A. T. & S. F. which, incidentally has remedied one fault found by many in the Code rule. The Santa Fe version reads:

15. The explosion of two torpedoes is a signal to proceed prepared to stop short of train, obstruction or switch not properly lined, for one mile. The explosion of one torpedo will indicate the same as two, but the use of two is required. The explosion of torpedoes must be acknowledged by Signal 14 (g).

This gets away from one objection offered to Code Rule 15, that it does not prescribe how far a train shall run at reduced speed after exploding torpedoes. One thing is certain, speed should be reduced until the train has covered the distance that a flagman would be expected to go back. With continually increasing speeds, the one mile would seem conservative enough.

Now let us take a look at the other school of thought: Those that believe an audible stop signal should be incorporated into the rules, in addition to a signal that requires reduced speed. Here is the Southern Pacific's rule:

> 15. The explosion of one torpedo is a signal to stop. When an unattended torpedo is exploded, train, after stopping, may then proceed with caution to a point not less than three-fourths mile from point where torpedo was exploded.
>
> The explosion of two torpedoes is a signal to proceed with caution for not less than three-fourths mile.
>
> Torpedoes must not be placed on public crossings, and must not be placed near station buildings, nor on yard tracks, except in an emergency.

There is some merit to the contention that, during severe storms, an audible stop signal is essential to safety. There may possibly be only one employe to protect trains in both directions in the case of a washout, slide, burned-out bridge, or broken rail; and no other means of stopping trains.

Regardless of which method is prescribed for use on your road, there are a few precautions that should be taken in handling torpedoes. They should be placed on the rail on the engineman's side against a splice bar so that there is the least danger of their being knocked off without being exploded. If there is any doubt about torpedoes being impaired by rain or snow, it is well to place two torpedoes, one on each rail opposite each other. This will insure an explosion of at least one of them. They should not be placed together on the same rail, as the explosion might be violent enough to cause damage.

When motor cars used by roadway men explode torpedoes, or damage them, these should be replaced at once.

Operators should keep torpedoes on hand to use during storms, placing them from 500 to 1000 feet from the order signal.

Torpedoes should never be placed near stations nor crossings where persons are liable to be injured by them. Inexperienced brakemen, firemen and laborers should be warned not to stand near torpedoes after placing them, because the caps, or other parts, may cause injury when exploded.

A bad practice is sometimes developed by enginemen, who write a message, fasten it to a torpedo and throw it off when passing a station. In one instance, a dispatcher walked outside the office to watch a train pass. The engineer threw off a torpedo with a message wrapped around it. The torpedo hit the ground, bounced back under the wheel of the train and exploded, burying a piece of said torpedo in the dispatcher's midsection. No great injury resulted but had it struck him in a more vulnerable portion of his anatomy, it could have been quite serious. No one ever knew what information the message contained, for it disintegrated.

RULE 16

Enginemen should comply with the communicating whistle signals and not jump to conclusions. Two short blasts when standing are a signal to start; but after the train has moved six inches, two short blasts are a signal to stop at once. Five blasts are a signal to increase speed;

and he should never take it for granted that he received the second two blasts because the conductor had not noticed the train was moving and repeated it. A passenger may have fallen under the train.

The two-air signal when running does not require an emergency stop but it does mean make a service stop "at once." And it should be answered by two short blasts of the engine whistle.

The air signal to back the train should be answered by three short blasts of the engine whistle as a warning to others that the train is to back; and a man, usually the conductor, must be on the rear platform of the passenger train. It is an unexpected movement and requires the exercise of care to avoid striking persons or vehicles, especially at crossings.

The air signal to stop at the next passenger station is answered by three short blasts of the engine whistle. Conductors should make this signal as early as practicable to give the engineman ample time to calculate the distance necessary for making the stop. Trainmen failing to get an acknowledgement of their air signals should repeat them, allowing ample time between pulls to give the air time to act. Two to four seconds is required, according to the length of the train.

Signals given improperly should not be acknowledged, but the engineman should call to have them repeated.

When an air signal is received to reduce speed, it is usually answered by two short blasts of the engine whistle. The chief objection to such whistling is that a flagman, not seen by the engineman, may have been giving stop signals and mistakes the engineman's answer as an acknowledgment of his stop; in which case he might stop giving stop signals.

If signaled to increase speed, or increase train heat, this is sometimes answered, although not really necessary.

When signaled to look back for hand-signals, the engineman should not answer until the hand signal is received.

The answer to a signal to recall the flagman is to call him in.

RULE 17

With the development of such safety devices as the gyrating headlight, it has been necessary to revise the rules in order to give them their greatest safety factor. The following Santa Fe rule serves to illustrate:

17. The headlight will be displayed to the front of every train by night. It must be extinguished when a train turns out to meet another train and has stopped clear of main track.

    It must be dimmed—

    while standing at stations where yard engines are employed; at meeting points by a train on the main track after switch has been lined for opposing train to enter siding; while standing to meet a train at a junction or at end of two or more tracks.

    On two or more tracks, it may be dimmed when approaching a train in the opposite direction.

    When headlight is displayed at meeting point, except within automatic block system limits or where Rule 261 is in effect, opposing train must proceed prepared to stop short of train, obstruction or switch not properly lined, until main track is known to be clear.

    On engines equipped with gyrating lights, the white beam will be displayed in addition to the headlight. When a train is stopped suddenly, by day or night, the red beam must be immediately

> displayed and trains on adjacent tracks observing red beam will stop and not proceed until it is known that their track is clear. Display of the red beam does not relieve employes from complying with Rules 99 and 102.

While this constitutes an improvement over Code Rule 17, there is still a hazard involved in extinguishing a headlight. A side collision recently occurred when an engineman "thought" he saw a green switch light at the rear of his train, assuming the train was in the clear with the switch lined for the main track. His train was not in the clear. It is difficult for an engineman to determine whether or not his train is clear of the main track. Capacities of sidings are based on an average car length of about forty-eight and one-half feet. With the advent of longer and longer cars in recent years, a train of seventy-five cars may not be able to clear in a siding whose capacity is listed at ninety-five cars.

Curvature frequently makes it impossible for the engineman to get a signal from the rear of the train.

Perhaps two-way radio, which is being adopted by some carriers, will solve this problem, eliminating the hazard involved in extinguishing the headlight.

There is another angle to the headlight business, too. This is probably the only rule which sets up a condition where a burned-out light bulb can cause a collision. If a light bulb burns out in the headlight at a critical moment when the train is pulling into the siding, the opposing train accepts this as a signal that the main track is clear. Accidents from such are a matter of record.

If a bulb burns out in a light signal, the signal must be regarded as displaying its most restrictive indication. No one would think of having trains regard a dark light signal as a "proceed" signal; yet this is precisely what Rule 17 indicates in the case of the headlight.

The rules of most roads require that, in addition to displaying the headlight at night and during the day when the view is obscured, the headlight will be displayed in tunnels and snowsheds at all times. Some roads have even gone so far as to require that the headlight be displayed both day and night. This is no doubt an added safety factor, at least at highway crossings.

Unless the engineman can determine positively that his train is in the clear, the headlight must not be extinguished. Also, before concealing it, he should note the position of the switch at the front of the train, if near it.

When two trains go to the same siding to meet a superior train, the following train must not assume that the leading train will protect it by keeping its headlight displayed until it is known that both trains are in the clear. The flagman should always protect the rear of his train against possible collision under such circumstances, not depending on the headlight of his, or any other train, for such protection.

The headlight of a train should never be extinguished until AFTER the train has stopped. To do so might result in the engineman of the train on the siding being blinded by the headlight of the approaching train, in which case he might run past the fouling point and collide with the approaching train.

The headlight of a train standing on the main track to meet another should be dimmed AFTER the switch has been lined for the opposing train to enter the siding. This practice has prevented collisions when the engineman of the approaching train miscalculated the distance or lost control of his train for some reason.

It must be understood, however, that the engineman of the approaching train is in no way relieved of his responsibility for a collision if the switch is not lined for him.

The rules of most roads require that, after lining a switch, the trainman take a position some distance from that switch, preferably on the opposite side of the track from the switch stand. This eliminates the possibility of his later becoming confused and throwing the switch back right ahead of the train. A good many collisions have been caused by such action.

When there are cars on a siding being used by a train, the engineman should always hold back where the cars will not conceal the headlight until after proper flag protection has been provided to the front of the train.

With the development of the oscillating white and red headlights, adopted on many roads, new rules have been written to govern their use. To illustrate, Rule 17 is shown below in its entirety, as found in the Uniform Code of Operating Rules, recently adopted by the Missouri Pacific and other roads:

17. HEADLIGHTS.—The standard white headlight must be displayed brightly to the front of every train by day and night.

 When a train turns out to meet another train, the standard headlight must be kept burning brightly until entire train is clear of main track; it will be dimmed while train is moving on siding entirely clear of main track, and must be extinguished when train has stopped entirely clear of main track.

 An extinguished headlight does not relieve train on main track from complying with Rule S-89 (a).

 It must be dimmed (except when approaching public crossings at grade):

 (1) Approaching and passing head end and rear end of trains, and engines standing or moving on adjacent tracks.

 (2) Approaching signals indicating train orders.

(3) When standing on main track awaiting arrival of an opposing train.

(4) When moving on sidings and other than main tracks, in clear of main track.

Outside of ABS territory, when headlight is displayed by train on siding at meeting point, opposing train must proceed at restricted speed until main track is seen to be clear.

"Winking" or "blinking" of headlights for any purpose is prohibited.

17 (a). WHITE OSCILLATING HEADLIGHT.—On engines equipped with a white oscillating headlight:

It must be displayed at night and during the day when weather conditions impair visibility.

It must be extinguished when the standard white headlight is dimmed or extinguished.

17 (b). RED OSCILLATING HEADLIGHT.—On engines equipped with a red oscillating headlight, such headlight will be displayed by day or by night when train is stopped suddenly under circumstances in which adjacent tracks may be fouled, or when head end protection is required.

The red oscillating headlight must be extinguished when necessity no longer exists.

A headlight burning red is a signal for an approaching train on the same or adjacent track to stop before passing such headlight, and be governed by conditions.

17 (c). NOT RELIEVED FROM PROTECTION.—Display of headlights does not relieve enginemen or trainmen from protecting train in accordance with Rule 99.

17 (d). RUNNING BACKWARD—WHITE LIGHT.—

When an engine is running backward, a white light must be displayed by night on the leading end.

17 (e). HEADLIGHT FAILURE.—When at night the standard white headlight fails en route, a white light must be used in its place, speed reduced, the bell rung continuously, the whistle or horn sounded frequently and train dispatcher notified.

17 (f). HEADLIGHT YARD ENGINES.—Yard engines will display standard white headlight to the front and rear by night. When not provided with a headlight at the rear, a white light must be displayed. The headlight may be extinguished on the end coupled to cars.

Road engines without cars standing or moving on other than a main track, or on main track within yard limits, must display a white light on leading and trailing ends.

RULES 18 and 19

Rule 18 calls for no explanation. A yard engine moves backward about as much as it does forward and needs a light on either end for obvious reasons.

With the coming of more and more diesel-powered streamlined passenger trains, Code Rule 19 will no doubt have to be somewhat altered to provide for the use of built-in markers, with which these trains are usually equipped.

In territory where trains are handled by signal control, such as CTC, which has been officially named "traffic control," the use of markers is sometimes eliminated as unnecessary. However, outside this territory markers are most important in railroad operation. Their absence from a train being met is an indication that the entire train has not arrived. If this fact is not noted, a collision might occur with the rear end of the train, which may have broken in two.

Theoretically, a break-in-two is not possible, as the train would go into emergency when the train line parts. Actually, as stated before, the capacity of air pumps has been increased to a point where the head end may proceed, the pumps keeping sufficient pressure in the train line to forestall emergency application of the brakes.

The experienced engineer can detect from the increased activity of his air pump when the train line has been opened, either from the caboose, or by the train becoming parted. However, if an inexperienced engineer does not detect the increased activity of the pumps, the partial setting of the brakes on the rear portion of the train may just about compensate for the loss of tonnage on his rear end. Such cases are a matter of record. Therefore, there is still a very real need to check closely on the markers of a train.

Helper engines, under unusual circumstances, have been known to cut loose from a train they have helped and run light downhill ahead of the train, depending solely on the absence of markers to protect the train. Under such circumstances, they are obliged to notify trains met that the rear of the train is following.

However, failure to do so would not excuse the opposing train in the event they missed the absence of markers and moved against the rear portion.

Where a bad-order car is handled behind the caboose, some kind of marker should be displayed on the car. Generally, a red flag is displayed by day and a red lamp by night on the rear corner of the car on engineman's side, and the markers are not removed from the caboose. There being no brackets on freight cars, some such plan must be followed. If air is not connected through, either the car must be chained to caboose, or a trainman should ride it.

Night markers, like the headlight, should be displayed while passing through long tunnels.

Markers should be removed when a run is completed.

The development and adoption on many lines of the oscillating red rear end lights is, like the gyrating white and red headlight, bringing rules to give these signals exact meaning. They provide an added safety feature, but have no bearing on the question of markers. Here is one recently adopted rule:

OSCILLATING RED REAR END LIGHTS—An oscillating red light displayed on rear of train is a signal for following trains on the same track to stop and for trains moving in same direction on adjacent track to move at restricted speed.

It must be displayed when train is stopped or fouling main track, or when train is moving under circumstances in which it may be overtaken.

Display of this signal does not relieve flagman from performing as prescribed by Rule 99.

—Uniform Code.

RULES 20, 21, 22 and 23

Signals should not be displayed on an engine until after authority is received to run; and they should be taken down promptly at the end of the run. The exception is where the train is to continue through carrying the same signals in charge of another crew. Then the signals may be left up and turned over to the relieving crew, provided they take charge promptly.

Care should be exercised to assure the proper display of signals. A dirty white flag and a dirty green one look so much alike that one may be mistaken for the other. The color of both is "dirty."

Accidents have resulted from the absence of the classification flag, or light from one side of an engine. Every effort should be made to assure both flags and lights being displayed. Looking at an engine from one side without signals, it is not difficult to miss the signal displayed on the opposite side.

The Code Rule 22 no longer requires that all engines display these signals, only the leading engine. However, many roads still prescribe that each engine shall display the signals. Others require that only the lead engine display signals, except where there is a short helper district and the helper is on the point, both engines will display them.

Where each engine is required to display signals, "coupled" means two or more engines coupled to each other. If cars are placed between any of the engines, the engine back of the cars would not display signals.

The green signals displayed for a following section are, at times, a deadly stop signal. An inferior train may have no knowledge that more than one section is being run on a schedule; and only the green signals restrict the train for the following section.

Operators and others concerned must notify the dispatcher promptly of any failure to display signals properly, such as one flag or light, very dim lights, etc.

If a train displaying green signals for a following section finds it necessary to tie up at an intermediate station under the hours of service law, the signals must not be removed until the schedule is more than twelve hours late. If possible, the dispatcher should be contacted in order that all trains may be properly protected.

In operating sections on a schedule, there is always the possibility of a following schedule passing the leading section without knowing it. The engine of the advance section may be hidden by a building, or on some track other than that on which the train is standing. The responsibility is usually placed with the crew of the following section, although some responsibility unquestionably rests with the crew of the leading section for failure to comply with the requirements of Rule 85, the exchanging of orders, numbers and signals.

The Santa Fe has incorporated a good rule in their book, reading:

85 (A). Unless otherwise provided, a section of a schedule or a superior train, which leaves the main track must not permit a following section or an inferior train to pass.

Where such a rule does not exist, extreme care should be used by the following section that they do not pass the train carrying their authority to proceed. If in doubt, they should stop and inquire the identity of a train at a station.

It is not good practice, but it does sometimes happen that a dispatcher restricts a leading section without giving a copy of the order to the following section. Therefore, if positions were reversed without the exchange of orders, a collision might result.

The importance of the engineman calling the attention to signals displayed by the use of the whistle signal cannot be overemphasized. If there is any doubt that the whistle signal is not understood, the train should stop and ascertain this fact.

Train indicators, where used, afford some protection for a second section should the green signals fail, or be missed by a train. However, they afford no protection for a third, or following section.

Should a train order refer to First 2 and No. 2 comes along without signals, either the crew of No. 2 should be questioned as to the absence of signals, or inquiry should be made of the dispatcher.

RULES 25 and 26

Rule 25 calls for little comment. Practically all passenger equipment in this country provides for the communicating signal appliance. In an emergency, it may sometimes become necessary to handle a passenger train with a freight engine not so equipped, which calls for extra precautions on the part of the crew handling the train.

The blue signal, provided for by Rule 26 for the protection of workmen, is generally well understood. Equipment must not be coupled into, or moved, when protected by the blue flag. The same workmen who placed the flag are the ONLY persons authorized to remove it. This may not be done by another person, even though the workman request that he do it.

When emergency work is necessary and no blue flag available, it is certainly only common sense to tell the enginemen what is to be done so as to insure no movement until the work is completed. Never assume that there is no reason for the train to be moved. The engineer might decide to back up a few feet to a water spout; or there are a thousand reasons why he might need to move the train to an extent that could cause injury to a person working between or underneath cars. Even the releasing of train brakes on a grade might prove sufficient to cause injury.

THE USE OF SIGNALS
Rules 27 to 35, inclusive

The rules of the majority of roads are much in line with Code Rule 27. When the light is out in a train-order signal at an open office on the Southern Pacific, they do not permit a train to be governed by the position of the arm, even though it can be plainly seen. The train must obtain a clearance. If the light is out in other semaphore signals, trains may be governed by the position of the arm.

In a straight color signal, there is no manner of determining the position of the signal. Should its most restrictive indication be yellow, or "caution," etc., then it would not be regarded as a "stop" signal. If the light is out in a signal capable of displaying a stop signal, then it must be regarded as displaying its most restrictive indication, "stop."

Imperfectly displayed signals should be reported promptly, so that steps may be taken to restore their normal indication. If a switch light is found displaying an imperfect indication, this can often be corrected by adjusting it.

The green-and-white signal used at flag stations should be so placed that it can be plainly seen. If the lights are placed too close together, the white light absorbs the rays of the green and the engineman cannot see the green light in time to make a proper stop. On single track, the signal should be placed on the office side, except where the view is obscured it may be placed on the opposite side. On two or more tracks, it should be placed alongside the track on which the train is running.

Where necessary to stop a train at a flag stop for any reason other than passengers, a stop signal must be used, not depending on the flag-stop signal, even though it be displayed.

Any object waved violently by a person on the right-of-way should be regarded as a stop signal. The person may have discovered some defect in the track which would make the movement of trains unsafe; and they may not understand the proper method of giving a stop signal.

Any signal not fully understood should be taken to mean danger; and trains should stop if necessary in order to make further investigation. A white rag, hat, or similar object on a stick between the rails should be regarded as a stop signal.

Any signal given to stop a train, other than a fixed signal, should be acknowledged by two short blasts of the engine whistle.

The use of the engine whistle is a matter requiring good judgment on the part of the engineman. The unnecessary use of the whistle is forbidden by rule. Where the use of the whistle in thickly populated areas is necessary, consideration should be given by the engineman to the nuisance angle. If it only be necessary for the whistle to be heard a short distance, then do not yank the whistle valve wide open—whistle softly. Any desire to whistle loud and long can be saved for a road crossing where it will do some good.

Crossing watchmen should be alert at all times and maintain a constant watch for potential hazards. Theirs is a responsible job, often performed under adverse weather conditions.

When a train or engine is moving, a constant lookout should be maintained in the direction of movement by both the fireman and engineer, when possible. The forward brakeman, too, should assist in this direction insofar as he is able.

There should be no failure to observe and call fixed signals before reaching them. Caboose men should also observe and call these signals, when possible.

There is an ever-increasing responsibility being placed on the men on the engine, for, as a rule, they are the only ones in a position to observe signals correctly. In Centralized Traffic Control territory, there is nothing between the engineer and a collision except his obedience to signal indication. He has no train orders; does not have the slightest notion where he may meet or pass a train, or be passed. On the other hand, he has no train order instructions to remember; neither does he have to keep his time-table constantly in hand so as to properly clear superior trains, although he must not get ahead of his schedule where time is shown in time-table. All of his instructions are conveyed by signal indication where, and when, he is to execute said instructions. The least he can do is observe, call and obey the signal indications religiously.

A green CTC signal turns red the moment the engine passes it; and the men in the caboose may not be able to determine whether the engine changed the signal or it changed just before the engine reached the circuit. Sometimes they are unable to see the signal at all until the caboose approaches it, when it is normally red and has no meaning.

There is no mystery about the flagging equipment prescribed by Rule 35, although there may be some question as to why engines are sometimes found minus this equipment at a time when it is badly needed. The equipment should be kept in good order and ready for instant use. There is always the question of how many torpedoes and fusees a flagman should carry. The Santa Fe has eliminated this question with a footnote:

Note: When going out to flag, flagman must take with him not less than 8 torpedoes and 6 fusees.

Chapter VII

SUPERIORITY OF TRAINS AND TRAIN MOVEMENTS

RULES S-71, D-71, S-72 and D-72

On single track, there are three ways in which one train may be superior to another, namely; Right, class, or direction.

On two or more tracks there is no superiority by direction; only right and class.

The sentence which confers on the train dispatcher the authority to regulate the movement of train is one sentence: RIGHT IS SUPERIOR TO CLASS OR DIRECTION.

Right is conferred by train order only, and is the tool of the dispatcher wielded by train orders. No orders, no right. Superiority can be conferred by time-table; but a train never has time-table rights.

Neither does a train ever possess ANY authority to proceed that the train dispatcher cannot take away from it at will.

Before a dispatcher can confer right on any train, he must of necessity restrict the precedence of some other train, or trains. He can take away from a regular train its superiority and confer said authority on an extra train, like this: "Extra 2525 East has right over No. 243 A to H."

Extra trains are inferior to all regular trains unless made superior by "right." There is no superiority between extra trains. If the dispatcher starts an extra from A to H, and another one from H to A, and does not make some provision for the two trains to meet, they will bump unless Providence intervenes. It is a lap of authority.

At a meeting point between extra trains, the train moving in the superior time-table direction will hold the main track; except that a work extra always takes the siding for any extra except another work extra. The rules of some roads provide: "At a meeting point between an extra and a work extra, the work extra must take the siding unless otherwise provided." If the rule book fails to set forth this fact, the work extra still has to take siding for an extra under an A. of A. R. decision which states: "Work extras must give way to all trains as promptly as practicable," puts the work extra in the siding.

Referring to the definition of extra trains, we find that the term "westward extra trains" includes all extra trains moving westward, including a work extra. But, if an order says "Westward Extras," that is something else again. Under the A. of A. R. Code, a passenger extra does not come in this category; neither does a work extra. A passenger extra may have right over all westward trains and an extra west may have right over eastward extras without creating any lap of authority. The westward extra would clear the passenger extra east and have right over eastward extras, which term does not include passenger extras and work extras.

The rules of all roads do not contain a definition of "passenger extra"; and the student is cautioned to examine his rule book. Where "passenger extra" is left out of the definition, orders such as those mentioned above would constitute a lap of authority; for the term "eastward extras" would include the passenger extra, as well as all other extras except work extras.

This point will be illustrated later.

It must always be remembered there is not superiority between extra trains except that conferred by train order. An extra in the superior time-table direction holds the main track at a meeting point; however, it is in no way superior to the westward extra. An order for two extras to meet is a restricting order to both trains.

And remember, the train dispatcher can take away any authority which any train holds to proceed on the main track. Sometimes "The time-table giveth, but the dispatcher taketh away."

RULE 82

This rule is worthy of considerable thought. A regular train, which becomes more than twelve hours late on either its arriving or leaving time, loses both right and schedule; and can then proceed only as directed by train order. The question is: "Exactly when is a train more than twelve hours late?"

It has long been a commonly accepted practice, that if No. 2 is due by Station F at 10:01 p.m. and the train can pass that point at 9:59 a.m., it has a perfect right to proceed on its schedule. Or, even if it can pass at exactly 10:01 a.m., it is still "Not more than twelve hours late on its schedule."

An opposing train waiting at G for No. 2, could not proceed until after said train became more than twelve hours late on its leaving time at G, say 10:15 a.m. Bear these figures in mind and we will come back to them presently.

If a train arrives at a station under flag, having died on its schedule, the dispatcher can put the train back on its schedule, provided there is dead time between the arriving and leaving times, say as much as one hour.

Even though the schedule has expired on its arriving time, it is still in effect until it becomes more than twelve hours late on the leaving time.

Should the train manage to scratch into such a station a minute or two before it becomes more than twelve hours late on its arriving time, it does not lose its schedule and may depart if it can do so before it becomes more than twelve hours late on the leaving time.

However, if a train once loses its authority under this rule, it is in the hands of the dispatcher. It cannot arrive one minute more than twelve hours late and retain the schedule, even if it can depart before the schedule expires on the leaving time.

This has been the commonly accepted interpretation for more years than the writer can remember. However, to show how even the most stable and well-established rule is often given an interpretation by some road, let us look at the Southern Pacific's Rule 82. It reads:

82. Time-table schedules, unless fulfilled, are in effect for twelve hours after their time at each station.

Regular trains more than twelve hours behind either their schedule arriving or leaving time at any station lose both right and schedule, and can thereafter proceed only as authorized by train order.

82 (A). Regular trains, unless otherwise provided, will be authorized at their initial stations by clearance, which must bear the OK, time, and initials of the chief train dispatcher. At an intermediate station, the following form of train order must be used:

"ENG. . . . . RUN AS NO. . . . . FROM . . . ."

The wording of Rule 82 is identical with the Standard Code and the rules of every other known road. Yet, it has a meaning different from that known to exist anywhere else.

On the S. P., the train waiting at G for No. 2, could proceed from G at 10:02 a.m., or when No. 2 became more than 12 hours late on its schedule at F. They set forth the fact that a schedule train must be due at every point on the railroad at some time; that is, there must be a due time on every foot of track.

When No. 2 is due to leave F, they are due to arrive at G; the leaving time at F being projected, so to speak, to constitute an arriving time at G.

Now, if No. 2 cannot pass F in sufficient time to reach G within this projected arriving time, 10:01 a.m., they are ruled dead on their schedule.

In other words, the engineer must figure his running time from F to G and deduct it from his leaving time at F to determine whether he has time to go—whether or not he is dead on his schedule.

This case is cited, not for the purpose of controversy, but to show why each student must delve into the rules of his particular road. It is impossible for a book of this kind to cover all similar rulings on all roads; therefore, the book cannot constitute an authority as far as any one individual road is concerned.

The safety feature which possibly eliminates the hazard from such a ruling lies in the Southern Pacific's instructions to dispatchers. They are instructed to take a train off its schedule before they get anywhere near twelve hours late.

Operators must keep in mind that an order addressed to a regular train is in effect for it until it becomes more than twelve hours late on its leaving time. It must not be assumed that a train is dead on its arriving time, if two times be shown. The train may have arrived without his knowledge.

Some roads permit operators to file orders addressed to trains which have lost their authority as prescribed by Rule 82. Other roads insist that these orders be annulled.

RULE 83

Shown below are some of the additions which various roads have made to Code Rule 83.

83 (A). The information called for by a train register

must be inscribed properly therein and the register checked by the conductor, or engineer if there is no conductor, except as hereinafter provided.

An extra train will register only at a register stationwhere it originates or terminates, unless otherwise directed.—Southern Pacific.

83 (B). A train may leave a register ticket of prescribed form with the train-order operator at a train-register station when authorized by special instructions in the time-table, or by train order, or when a train-order check of trains, Form V, is received. When so authorized, the operator must enter in the register information contained on the ticket, then report the train from the register.

When a train order check of a schedule or section, Form V, is received, or a schedule or section is checked on a register at the initial or terminal station, or at the end of double track when passing from double to single track, or after having been met on single track or passed on either double or single track by a regular train, it will not be necessary to check an intermediate register against the same train.—Southern Pacific.

Note: Special attention is called to the second paragraph of this rule inasmuch as it is representative of one of two schools of thought with regard to train registers, or their use. This subject will be discussed fully later.

83 (C). Before leaving a train-register station at which it is necessary for the conductor to check the register, he must deliver to the engineer a check of the register on prescribed form and retain a copy. —Southern Pacific.

83 (D). When a train is restricted for an extra train, the restricted train must not leave unless the extra train

has been identified by the conductor or engineer, or a train order is received superseding or annulling the restriction, or a train order Form V, Example

CHECK OF TRAIN REGISTER

By Conductor.............................*Train*....................

At.............................. ....*M*. *192*
(Station) (Time) (Date)

TRAIN	SIGNALS DISPLAYED	ARRIVED	LEFT
....	....	....	....
....	....	....	....
....	....	....	....
....	....	....	....
....	....	....	....
....	....	....	....
....	....	....	....
....	....	....	....
....	....	....	....
....	....	....	....
....	....	....	....
....	....	....	....

Before leaving a train register station where it is necessary to check train register, conductors must be governed by Rules 83(A), 83(B) and 83(C) and enginemen must require from them a check on this form of the numbers of superior trains and trains of the same class that have arrived or left.

(3), is received. (EXTRA ...... HAS ARRIVED (or left) ...... ON ORDER NO. ....).

A train may check the register against an extra when authorized by train order Form W, Example (4). (........ MAY CHECK REGISTER AT ...... AGAINST EXTRA ......ON ORDER NO. ....).—Southern Pacific.

83. (Second paragraph): A train that has been detoured or that has had its identity temporarily changed by train order, will not resume its original schedule unless so authorized by train order.—Illinois Central.

Note: This Illinois Central rule is essentially the same as the Southern Pacific's, a reversal of the practice on many roads, as authorized by the A. A. R.

A conductor should never register out of a terminal until he has received a clearance card. Otherwise, he may write in the wrong train or signal information and overlook correcting the record.

Operators must OS trains from the train register, always reporting whether "no signals" or "green signals." Should an error be made in registering, the dispatcher is then in a position to have the record corrected.

Seeing a train at a register station without classification signals does not relieve a conductor from checking the register. The train may have arrived with green signals and taken them down prior to his arrival.

When a train is restricted for an extra, it is necessary to actually see the train or have it identified by the conductor, as it cannot be checked on the register. The exception is on roads which permit the dispatcher to issue a train order authorizing a train to check a register against an extra.

Where helper districts exist, it is not unusual for an

engine to make several trips as an extra in one day; however, the train dispatcher is in a position to know when confusion may arise. The Southern Pacific's order authorizing a train to check a register against an extra puts the responsibility for hazard right in the dispatcher's lap. This order is quite useful in cases where an extra terminates at a blind siding.

A train-order check of trains at a terminal may eliminate the necessity for the conductor checking the register and filling in the form for delivery to the engineer; however, it is still his responsibility to check against schedules which may have become due after the time specified in the order.

Some roads, such as the Southern Pacific, have eliminated intermediate registers, except for trains which originate or terminate at intermediate points. The display of signals to an intermediate point is forbidden; and where a section originates at other than the initial station of the schedule, a copy of the order creating the section must be given all trains affected to protect the movement. Likewise, signals cannot be taken down at an intermediate point. If a last section is to be discontinued at an intermediate point, signals for it must be displayed through to the terminal and the section annulled, using Form K.

Under such an arrangement it is perfectly safe to require that a schedule, or section of a schedule be checked only once.

A regular train is usually authorized from its initial station by a clearance card. Where a train is to be put on a schedule at an intermediate point, a clearance card is sometimes not sufficient. There is considerable controversy regarding the proper manner in which a train, having been detoured, is to assume its former identity upon returning to its own line. The A. of A. R. has ruled as follows:

Is it necessary for a detoured train to have a train order before resuming its schedule on returning to its own line, for instance: A train scheduled from A to Z detours from E to N. May it resume its schedule at N without a train order instructing it to do so?

Assuming that Form K has been used to annul the train between E and N, there is nothing to prevent it resuming its schedule N to Z without a train order. (April 19, 1909).

Many carriers still adhere to the AAR ruling; the reasoning being that once authorized to fulfill a schedule, it retains such authority until the schedule is fulfilled, annulled, or becomes more than twelve hours late. It is usually required that the crew, upon returning to its line, report to the train-order office to see if there are instructions; however, if there be no train-order office at that point, they may proceed without clearance or any further authority to reassume their schedule.

Where the train is detoured from its initial station to some intermediate point and never was duly authorized to fulfill the schedule at its initial station, it must then receive train order authority to assume the schedule upon returning to its line. A schedule cannot be assumed at any point other than its initial station without a train order to do so. At the initial station, of course, a clearance card is all that is required.

Because of the many complications which have arisen out of the foregoing practice, creating hazard when a train which had been detoured showed up unexpectedly, often at a point not an open train order office, assuming a schedule in such a manner as to render it beyond the control of the train dispatcher, some roads have set aside the A. of A. R. ruling and incorporated into their books a rule similar to the following:

"A train that has been detoured, or that has had its iden-

tity temporarily changed by train order, will not resume its original schedule unless so authorized by train order."—Illinois Central.

Under such a rule, a train taken off a schedule for any reason, whether detoured or not, cannot again assume the schedule, regardless of whether the schedule, or a part of the schedule has been annulled or not. If a train is run on a schedule existing between A and Z; and at D the dispatcher annuls the schedule D to H and runs the train extra between those points; the train cannot again assume the schedule at H without a train order to do so.

RULES 84 and 85

The signal to leave a station must be authorized by the conductor. It may be passed to the engineman by another member of the crew.

A signal to leave must not be accepted by an engineman unless it is known that superior trains can be cleared at the next station. By night, the engineman must know positively that a signal was intended for him before leaving, when in a yard where there are other trains, or yard engines.

A proceed signal given from the rear of a long train should be given with a white flag, as the arm cannot be readily seen. Such a signal should be answered by two long blasts of the whistle where no whistling restrictions prevent.

It is never safe to give an approaching train a proceed signal because the crew at the station has no way of knowing what restrictions the approaching train may hold. When the engineman discovers the train signaling him is not the one he has orders to meet, it may then be too late to avoid an accident.

The Santa Fe has added a paragraph to their Rule 84 which seems sensible. It reads:

> "A train must not be backed until proper signal has been received from the rear end."

Because Code Rule 85 is partially blank, here is the completed rule as taken from the S. P. rule book:

85. When a train of one schedule is on the time of another schedule of the same class in the same direction, it will proceed on its own schedule.

    Trains of one schedule may pass trains of another schedule of the same class. Second and inferior class and extra trains may pass and run ahead of second and inferior class and extra trains.

    A section may pass and run ahead of another section of the same schedule, first exchanging train orders, signals, and numbers, with the section passed. The change in sections must be reported from the first available point of communication.

Just why the S. P. still retains the first paragraph of Rule 85, which was dropped from the Standard Code in the 1928 revision, is not clear. It means, of course, that while the time-table may show a passing point for two trains of the same class, it is not necessary for the train to be passed to wait indefinitely at the passing point. It may proceed on its own schedule. This paragraph will not be found in a great many rule books.

There is at least one road in this country which has not yet adopted the second paragraph of Rule 85. Extra trains must clear the time of all regular trains in the same direction. The dispatcher must give a third-class local a run-ahead of a second-class train, etc.

On the other hand, the majority of roads do not require that second and inferior class, and extra trains clear anything except first-class schedules. They may run ahead, or pass each other as conditions require; and the dispatcher's time is not taken up putting out run-ahead orders to keep an extra train ahead of the local. Chances are, neither of them is doing a job of getting over the road, anyway.

One section should not pass another without train-order authority, when such authority can be had; for instance, at an open train-order office. Care must be exercised in the exchange of train orders, numbers and signals. Dispatchers should deliver copies of all train orders to all sections liable to be affected by such an exchange; however, this is not always done. The exchange should be reported from the first available point of communication so the dispatcher can straighten out the mess and word his train orders properly thereafter.

RULES 86, 87, 88 and 89

Rules 86, 87 and 88 will be found considerably amended in most rule books, if not actually scrambled. Here is the Santa Fe version:

86. Unless otherwise provided, an inferior train must be clear at the time a superior train in the same direction is due to leave the next station in the rear where time is shown, but not less than five minutes; except, within automatic block system limits an inferior train must clear the time of a superior train in the same direction only sufficiently to avoid delay to the superior train.

    Unless otherwise provided, an engine must clear the time of a first-class train at the time the first-class train is due to leave the next station where time is shown, but not less than five minutes; except, within automatic block system limits an engine must clear the time of a first-class train sufficiently to avoid delay to the approaching train.

S-87. An inferior train must keep out of the way of opposing superior trains and failing to clear the main track by the time required by rule must be protected as prescribed by Rule 99.

Extra trains must clear the time of opposing regular trains not less than five minutes unless otherwise provided, and will be governed by train orders with respect to opposing extra trains.

88. Trains must pull into the siding when practicable; if necessary to back in, or back out, the train must first be protected as prescribed by Rule 99, unless otherwise provided.

Now let us take a look at the S.P. version:

86. CLEARANCE OF TRAINS WITHIN BLOCK SYSTEM LIMITS:

(a) At meeting points the inferior train must clear the main track before the leaving time of the superior train.

(b) When arriving time is designated for a first-class train, or a train of superior right, an inferior train in the same direction must clear that time sufficiently to avoid delay to the superior train.

87. CLEARANCE OF TRAINS OUTSIDE OF BLOCK SYSTEM LIMITS:

(a) At a meeting point between trains of the same class, the inferior train must clear the main track before the leaving time of the superior train.

(b) At a meeting point between trains of different classes, the inferior train must clear the main track not less than five minutes before the leaving time of the superior train.

(c) Extra trains must clear the time of opposing superior trains not less than five minutes.

(d) An inferior train must clear the time of a first-class train or a train of superior right in the same direction, before the superior train is due to leave the next station in the rear where time is shown, but not less than ten minutes.

S-88. Extra trains will be governed by train orders with respect to opposing extra trains.

At a meeting point between extras or between work extras, the train in the inferior time-table direction must take the siding, unless otherwise provided.

At a meeting point between an extra and a work extra, the work extra must take the siding, unless otherwise provided.

88 (A). Firemen must remind engineers, and when practicable brakemen must remind conductors or engineers, of the time of a superior train which must be cleared.

The provision that an inferior train must be clear at the time a first-class train is due to leave the next station in the rear where time is shown, is based on the theory that when No. 2 is due to leave F, it is due to arrive at G. Some roads require only a five-minute clearance when this time between stations is less than five minutes, like the Santa Fe. Others, like the S. P., require a ten-minute clearance, if the running time be less than ten minutes between the two points.

The S. P. Rule 87 (d) includes "trains of superior right" because they still retain the old Form G (3) order, commonly miscalled the "schedule order"; and they require a train running on such authority to be cleared in the same manner as first-class trains. The old Form G (3) order has been dropped from the Standard Code; and a good many recent rule books have done likewise. Therefore, extra trains cannot be given "right" over train moving in the same direction and there is no need to include "trains of superior right" in their rule.

Considering the Form G (3) order, it might well seem that their Rule S-88 should read: "Extra trains will be governed by train orders with respect to other extra trains," instead of "opposing extra trains."

When a train is running against the schedule of a superior train of the same class, a one-minute clearance is usually required. This is based on the permissible 30-second variation in watches. If one watch was 30 seconds fast, the other 30 seconds slow, a collision could result, except for the one-minute requirement.

If a train is unavoidably delayed between stations and it becomes apparent that they have not sufficient time to take their train and make the next station in advance and clear as required by rule, then the engine should be cut off and run light to the station in advance to hold the superior train. It is never safe to use any portion of your clearance time in such a case.

An accident was recently reported by the Interstate Commerce Commission where a head-end collision resulted from an engineer failing to get a flagman out ahead when his engine steamed badly. He was hit just 51 feet short of the siding switch where he was to head in and a few seconds after his wait had expired. He gambled on the train not being on its wait order—and he lost his life.

Should a second or inferior class or extra train be going to C for No. 2, they must clear No. 2's leaving time at least five minutes. "In the clear" means clear on the siding with the switch lined up properly for the main track.

If an extra train is given right over an opposing extra

and directed to wait at a certain station, the extra over which right is given must clear these wait times at least five minutes the same as it would for an opposing regular train. In creating extra trains, dispatchers should get in the habit of entering an extra train on the train sheet before giving them a running order. If an extra is to run over an entire subdivision, it is not good practice for the dispatcher to give them a running order over only a part of the subdivision as a matter of convenience. The dispatcher must always remember that, having created an extra, its movement must be protected against any opposing extra movement that he may authorize. Work extras should not only be endorsed on the train sheet in ink, but their limits should be marked by a pencil notation in such a manner that they will not be overlooked. By marking a diagonal line across the station column on the train sheet, showing the limits of the work extra, this is a constant reminder not easily overlooked. In connection with observing proper clearance, consideration should be given to view, weather, length of siding at meeting stations, grade, speed of opposing train and whether or not it is scheduled to stop. More leeway must be allowed in some cases than in others. The inferior train must pull into the siding at the nearest end and not go to a crossover or to the other end of the siding to back in, unless the movement is fully protected, as per Rule 99.

Since the opposing superior train waits for its time at the point where an opposing inferior train would clear, it will be seen that after the superior train is due to leave the next station in advance, it then becomes due all the way between that station and the point where the inferior train should pull into the siding.

Should No. 16 be instructed by train order to take siding and meet No. 15 at a point where there is a spur track only, as shown by Chart A, it would seem to be unnecessary for No. 16 to flag while pulling by to back in, as it must be known to the crew of No. 15 that there

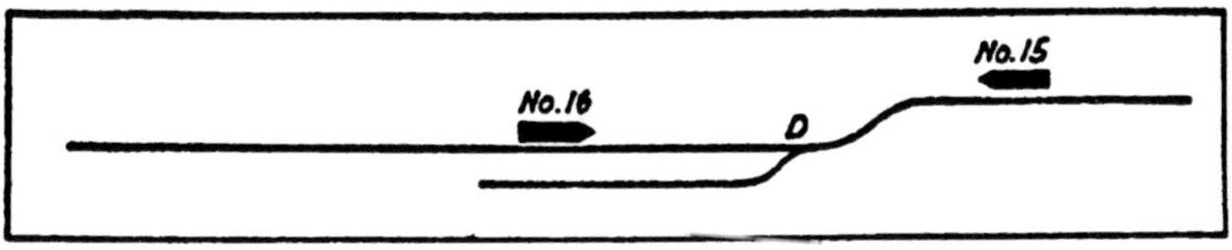

is no other way for No. 16 to take siding; but in the absence of train order, in case No. 16 is required by time-table to take siding at D for No. 15, it will be seen that a flagman must be sent out, as No. 15's time applies where No. 16 is required to go on the siding.

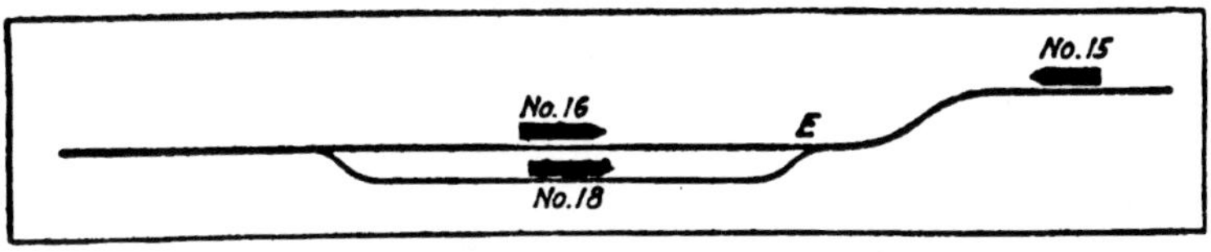

Should No. 16 be instructed by train order to pull by and back in at station E, Chart B, responsibility for collision would rest with the crew of No. 15 if given the order before arrival at E; however, under the rule, and as it is applied on many roads, still No. 16 must send a flagman in advance from E to a point well beyond the territory necessary for it to use in pulling up to back in. When a dispatcher fixes a meeting point in such cases he should indicate in the order the number of cars in No. 16's train. No. 16 must leave a flagman, at the point where its engine stops to back in, to inform No. 15 that it has arrived and is backing in. No. 15 must stand back far enough to allow No. 16 room to back in, and remain there until No. 16's flagman advises that his train has arrived. No. 15 must not go down to the station when a train is seen on the siding, as it may not be No. 16, but No. 18, some extra, or a work train instead.

While there is no particular objection to a superior train taking siding when it is entitled to hold the main track, if this will expedite traffic, it sometimes happens that confusion arises where the view is obstructed and the inferior train enters the other end of the siding. It may be necessary at times for an inferior train which should pull into the siding to hold the main track against a superior train under protection of flag. This usually is the result of having held the main track at a meeting point and then being delayed until there is insufficient time to go to the next siding for a superior train.

However, this should not be done when it can be avoided.

On one road, it became general practice for heavy freight trains to pull up the main track and put passenger trains through the siding at a meeting point. This continued for a considerable time until one night during a heavy fog, a freight train pulled up the main track between siding switches and failed to get its flagman out to stop a passenger train, with the result that the passenger train collided with the freight while moving at considerable speed.

As a rule, it is better for trains to go where they belong; that is, if they should take siding, they should go on the siding, and if they are entitled to the main track, they should hold the main track. This eliminates possibility of confusion and hazard.

Sending a restricting order to a train at the point where it is restricted requires considerable care on the part of the dispatcher. For instance, to issue an order to the effect that No. 15 would get the order and meet No. 16 at F instead of E in itself does not authorize No. 15, the train of inferior direction, to hold the main track. While it should be apparent to the crew of No. 16 that No. 15 would come down the main track at F before receiving the order, nevertheless there may be some misunderstanding as to whether No. 15 is actually entitled to hold the main track after receiving the order. The majority of roads require that the order read:

> No. 15 get this order hold main track and meet
> No. 16 at F instead of E

Another form sometimes found reads this way:

> No. 15 meet No. 16 at F instead of E
> Order to No. 15 at F
> No. 15 hold main track at F

The former order is shorter and just as clear at the latter.

Code Rule S-90 has been expanded on the majority of roads in the case of passenger trains. The conductor on a train of passenger equipment must first give one short blast on the communicating signal approaching a meeting or waiting point, 16 (L), to be answered by two long and one short blasts of the engine whistle, signal 14 (n). Below will be found the Southern Pacific's rule S-90 (A). The individual rule book must be checked for the communicating signal to be used approaching a meeting point, as this signal is not the same on all roads.

S-90 (A). Approaching a schedule meeting point with a train of the same class, or a point where by train order the train is to meet or wait for an opposing train, the engineer must sound signal 14 (n) immediately after sounding signal 14 (m). On a train of passenger equipment the conductor must also sound signal 16 (l) at least one mile before reaching the station. Should the engineer fail to sound signal 14 (n), fireman must immediately remind engineer of the requirement, and should engineer fail to prepare to stop short of fouling point, when required, the conductor must take immediate action to stop the train.

Should a stop be made after signal 14 (n) has been sounded and before the expected train is met, signal 12 (c) or 16 (a) must not be given for further movement until such train has been met, or authority is held to proceed against it.

RULES 91, 92 and 93

While some roads have adopted Code Rule 91 as given, the majority of roads still require a ten-minute block between all trains outside block-signal territory.

It will be noted that the 1938 revision of the Code dropped the first paragraph of old Rule 92, which read:

> "A train must not arrive at a station in advance of its schedule arriving time."

Many rule books still contain this paragraph; although, on some of the lines its provisions have been nullified by special instruction in the time-table, which permit a train to arrive before its schedule arriving time.

Since Code Rule 93 is in skeleton form, the rules of several roads will be given so as to give the student some idea of how the blank spaces are filled in on different carriers. It will be noted that some of the rules employ a wording not suggested by the Standard Code; however, these usually have about the same meaning as the rules of roads which simply fill in the blank spaces of Code Rule 93.

93. Within yard limits the main track may be used, protecting against first-class trains.

Second and inferior class trains, extra trains, and engines, must move with caution within yard limits.

When not protected by block signals or when moving against the current of traffic, first-class trains must move with caution within yard limits.

Note: Second and third paragraphs of this rule apply to all tracks within yard limits.—Southern Pacific.

93. Within yard limits (designated by yard limit signs), the main track may be used, clearing first-class trains at the time shown at the next station in direction of the approaching first-class train, except:

If the distance between stations is less than 3 miles, or if the time between stations is less than 5 minutes, a train or engine must be in the clear 5 minutes before the time shown for the first-class train at next station in direction of approaching first-class train.

If not clear by the time required, train or engine must be protected at that time as prescribed by Rule 99.

Within yard limits, the main track may be used without protecting against second- and inferior-class trains, extra trains and engines.

Within yard limits, second- and inferior-class trains, extra trains and engines must move at restricted speed.—Missouri Pacific.

93-A. Block signal indications within yard limits do not relieve trains and engines from moving at restricted speed as required by Rules 93 or D-93.—Missouri Pacific.

D-93. Within yard limits, when moving against the current of traffic, all trains and engines must move at restricted speed.—M. P.

93. Within yard limits, movements not authorized by time-table schedule or train order may be made on the main track by proper signal indication or permission of the operator without protecting against extra trains or engines.

Within yard limits, regular and extra trains are not required to provide protection, except against regular trains.

Extra trains and engines must move within yard limits prepared to stop short of other yard movements, trains and obstructions unless the main track is known to be clear.

Yard engines will not display markers except when passing block stations.

Note: Where block signal rules are in effect "known to be clear" includes when track is known to be clear by signal indication.

Note: Yard limits will be indicated by yard limit boards.

Note: Yard engines will display a red light to the rear at night.—Pennsylvania Railroad.

93. Stations having yard limits will be designated by special rule in time-table.

Within yard limits trains and engines may use main track not protecting against second-class and extra trains or engines, but must give way as soon as possible upon their approach.

All except first-class trains must move within yard limits prepared to stop short of train, obstruction or switch not properly lined; the responsibility for accident with respect to second-class and extra trains or engines rests with the approaching train or engine. First-class trains moving against the current of traffic must move within yard limits prepared to stop short of train, obstruction or switch not properly lined, except where Rule 261 is in effect.—Santa Fe.

93. All second- and third-class and extra trains must approach and pass all stations, yards, sidings, fuel and

> water stations, expecting to find trains or yard engines occupying the main track within the switches or yard limits, or taking fuel or water and will be prepared to stop unless the main track is seen or known to be clear. Trains and yard engines may use the main track at such points, protecting against first-class trains in all cases and will protect against all trains where view is obscured by fog, storms or track curvature, or other causes requiring additional safeguards.
>
> Yard engines must clear the time of first-class trains not less than five minutes, but must be clear at the time a first-class train is due to leave the first station where time is shown, east or west, of the location where yard engine is performing work.—Chicago & North Western.

Referring again to Rule 91, on many roads operators are required to space trains passing their stations by using the train order signal and clearance card for this purpose, clearance card specifying the time train may leave that point. When so held at a station, the train should pull down far enough to permit any opposing train met to leave the station. When an inferior train is given an order to run ahead, the following train should not depend entirely on a block of this type but should take into consideration the character and tonnage of the train preceding it. Caution should be exercised approaching sidings as the leading train may be required to go on a siding for an opposing train for which the following train is not restricted.

If the superior train is given an order that an inferior train will run ahead of it A to B, it is known that the inferior train must stop at the first switch at B to take siding. The superior rain should approach that switch carefully. While the leading train should not use the run-ahead order as an excuse for a poor job of flagging, there is little excuse for the following train to approach such point at high speed simply because they have not been flagged. "Run ahead" orders should be saved for those occasions which border on an emergency or when necessary to raise a blockade. If the move can be made by using a run late or time order, the run ahead order should never be used. Certainly, such an order should not be given at all during heavy fog or at night.

The various quotations from Rule 93 reflect the ideas of various carriers as to the proper manner in which to require trains and engines to move within yard limits. Examination of Interstate Commerce Commission reports reveal a startling number of accidents occurring within yard limits due to various rule violations. These accidents, coupled with collisions elsewhere due to faulty flagging account for a large percentage of the total number of accidents investigated.

If all yards were laid out scientifically or were kept within a limit of one or two miles, the problem of writing a yard rule would not be so difficult. Where the main track must be used almost constantly and the limits of the yard cover from three to ten miles, it will be seen how difficult it is to formulate a standard rule to meet the requirements of every road.

Taking into consideration the definition of "yard speed," the situation seems to be well taken care of by the foregoing rules. There is, of course, a difference of opinion among carriers as to whether first-class trains should be "cleared" rather than "protected against." Where trains and engines are required to clear first-class trains, they must be in the clear at such time as the first-class train is due to leave the first station in advance, and they must not again occupy the main track even under flag protection.

A rule such as the C. & N. W. rule quoted would be a great delayer of trains on roads traversing hilly country, where tonnage trains might have to "double" hills if speed were reduced at each siding as required. Considerable hazard can arise from such a rule if not enforced to the letter. On one road which was not equipped with block signals, the officials winked at speed violations, in restricted territorry, such as at sidings. Had a train complied with yard limit rules at each siding they would have been doubling hills all night. Trainmen and enginemen could keep track fairly well of the trains ahead and following during normal times; however, when the road was hit suddenly with a peak traffic load, not even the dispatchers could keep track of all their trains, much less trainmen and enginemen. Within a period of a few weeks there were three bad rear-end collisions at one water tank, and others elsewhere. Any potential hazard will ripen into manifestation if given sufficient time.

The object sought in Rule 93 is to enable yard crews, (or train crews where no yard engine is used) to do their work without having to keep flagmen out in one or both directions, when necessary to occupy the main track, as the main track is not a part of the yard in any sense whatever.

There should be no place on a railroad where a passenger flagman is relieved from protecting his train properly if in a position to be overtaken by a following train, yard engine, work train, or by hostlers handling road engines to and from trains.

All trains and engines which must run at restricted speed should move prepared to stop within one-half the distance seen to be clear.

Should an engine, car or train be struck by a train which is restricted by Rule 93, in its movements through yards the engineman of such train should, of course, be held responsible for the collision; but this does not mean that cars may be left on the main track by night or during rain, snow or fog, unprotected, nor does it imply that a crew may allow the rear of their train to stand on an obscured curve without protecting it in a reasonable and sensible manner. If your train or its markers cannot be seen plainly for at least 20 car lengths, protect it.

Regardless of where he may be working, no man is justified in permitting his train to stand during fog, or on a sharp, obscured curve, without protection, either in yards or elsewhere.

A train order to run ahead of a train does not relieve the crew from protecting within yard limits the same as outside of such limits, the only difference being that within yard limits sometimes it is not necessary to go back as far, since certain trains must reduce speed at such points.

Within yard limits is a very poor place to run at speed. Even passenger enginemen should run with care, according to the conditions as they present themselves at the different points.

As a train whose speed is not restricted by this rule may be following, and passenger flagmen have no means of ascertaining whether or not this is the case, a hostler may not be on the alert, or may be inexperienced; or yardmen may be shoving a string of cars ahead of engine without exercising due care. It will be seen that there never will be a time when a passenger train can be left standing without its rear being guarded.

If enginemen of either freight or passenger trains can see that the way is clear ahead for, say, one thousand feet, they have every reason to assume the track will remain clear for that distance and you must not shove cars out in their faces, or move against them. If you put enginemen in position where they cannot avoid striking your cars, you will be held responsible for the collision.

If a first-class or any other train or engine, except regularly assigned yard engines, is delayed or makes an unusual stop within yard limits it should be protected. "Unusual stop" must be understood to mean almost any stop on the main track.

When an extra train is given right over all trains on a Form G (3) order, it is not excused from complying fully with Rule 93. Cases have been known where such orders conferred right on an extra over yard engines as well as all trains. This is not possible under the rules.

A yard engine derives its authority to work directly from the rules; the movement is not created by the dispatcher and he cannot control the movement of the yard engine. Therefore, he cannot confer something on a train which he is unable to take away from a yard engine.

Yard engines must not run at a high rate of speed on the main track within yard limits, or otherwise totally disregard all safety precautions to prevent collision with trains, cars or other yard engines; especially during darkness, thick weather or where the view is obstructed. Extreme care must be exercised in moving cars over public crossings, or on tracks where car repairers may be at work. Notwithstanding the burden of responsibility for accident rests with certain approaching trains, yardmen must keep a proper lookout. They must not pass through a crossover from behind cars or other obstruction in such a manner as to make it impossible for an approaching train or engine to avoid striking them, when the crew of such approaching train or engine can see for some distance ahead and have reason to think the track is clear. The crew on the switch-engine would be responsible then for any accident that might result, provided the approaching train was properly under control.

To sum up: The only way in which to get the work done and at the same time get trains through yards without burdensome delays or accidents is for every employe to be on the alert, exercise reasonable judgment, move at a sensible speed and always put himself in the other man's place, asking himself what the other man can or should do, and what he cannot or should not be expected to do. Often it is practicable for yardmen to get information from the dispatcher as to about when trains may be expected. This is of great help to them in doing their work.

RULE 94

It will be noted that the second paragraph of Rule 94 was eliminated from the 1938 revised Code and many of the newer rule books have also dropped this second paragraph. With improved communications, Rule 94 has been eliminated from some rule books entirely; for instance, on the Southern Pacific. Other roads have retained Rule 94 as shown in the Code rules, while still other roads have yet in effect the old second paragraph which reads:

> "When a train, unable to proceed against the right or schedule of an opposing train, is overtaken between communicating stations by an inferior train or a train of the same class having right or schedule which permits it to proceed, the delayed train, after proper understanding with the following train, may precede it to the next available point of communication, where it must report to .... When opposing trains are met under these circumstances, it must be explained to them fully by the leading train that the expected train is following."

If a train becomes disabled at a point where the dispatcher can be contacted, the facts should be reported to him immediately. He can then make such arrangements as may be necessary to get the more important train moving.

Where there is an engine failure on an important train and it is overtaken by a train of lesser importance, it is only good judgment to take the engine of the following train and get the important train moving.

When asked as to what constituted a "disabled train," the Train Rules Committee of the A. of A. R. ruled:

> "Any trouble in regard to the meaning of the word 'disabled' could only occur when it is of minor importance. In such a case the judgment of conductors and enginemen must be used, always bearing in mind that the rule is intended to expedite the movement of trains."

After a train has exchanged with a disabled train between points of communication, it must report to the dispatcher as quickly as this is possible; that is, from the first point of communication reached. It then becomes his duty to issue such orders as may be necessary to protect the movement of both trains.

The second paragraph of old Rule 94 (for the benefit of those who work on roads still retaining it), has to do with a train that is being delayed because it is unable to move against the right or schedule of an opposing train. No. 244, let us say, is delayed at C waiting to meet No. 243 on a Form A order. Along comes Extra 4199 East holding an order to meet No. 243 at E, or two stations in advance. After a proper understanding with the crew of Extra 4199 East, No. 244 may precede the extra to E.

If D happens to be an open train-order office and the dispatcher attempts to restrict Extra 4199 East; it must be explained to him that No. 244 is preceding him, using his authority to move. Under these conditions, the dispatcher may not restrict the extra.

If Second 1 should be given right over No. 2 F to C; and on arrival at the schedule meeting point at C, No. 2

does not find First 1 there because of their having been delayed, the question arises:

May No. 2 leave C, or any point between C and F after the schedule of No. 1 becomes due, and First 1 has not arrived, and move against Second 1 until First 1 is met?

The answer is: NO. (A. A. R. June 24, 1926). Under the provisions of Rule 94, second paragraph, a section of a schedule may take another section of the same schedule ahead of it. Therefore, First 1 may be moving on such authority and a collision would result. This rule refers to a "Train," making no mention of the authority under which said train may be moving. Since "right" is superior to "class" and "direction," Second 1, and therefore First 1, holds authority to proceed to C against No. 2.

Such a situation could have been avoided by the dispatcher giving First 1, also, right over No. 2 between the points mentioned.

Chapter VIII

MOVEMENT OF TRAINS

RULES 95, 96, 97 and 98

It is interesting to note that the Pennsylvania Railroad has dropped Rule 95 from their book. Sections cannot be run on a schedule. One schedule, one train.

The many problems that arise out of operating trains as sections of a schedule will be set forth in the question and answer section in detail.

The first section of a schedule actually carries the authority on which the second section operates; and the second section the authority for the third, etc. Therefore, sections must maintain sequence. For the second section to run around the first without exchanging orders, numbers and signals, would almost certainly create a hazard. If an opposing train is given an order to move against a first section of a schedule, it moves against that section, disregarding following sections until such time as the first section is met.

A good many roads require that when a train dispatcher helps a train against one section of a schedule, an equal amount, or more help, must be given the train against all following sections. In other words, if Extra 47 East is given a meet with First 77, it must also have a meet with Second 77, or authority to proceed beyond the meeting point with First 77 for the second section. Although not practiced on many roads, such instructions constitute about the only safe manner in which inferior trains can be moved against sections without the possibility of hazard arising.

If a dispatcher gets in the habit of giving a train a meet only with a first section, when two or more sections are running close together and he wishes the inferior train to remain at the meeting point, this practice usually ends with his giving the inferior train an order to hold main track and meet the first section at some time. This places the train in the position of holding authority to the main track with the first section, when it does not hold authority to proceed beyond the first switch for the second section. If the inferior train does hold the main track between switches, it should do so only under adequate flag protection, not relying on the first section to hold the second from proceeding up the main track at that point.

The fact that a leading section has registered as leaving a station must not be taken as conclusive evidence that it has left the yard. Responsibility for knowing that it has actually left still rests with the following section.

The Code no longer authorizes the display of signals to or from an intermediate point of a schedule unless that point be a register station for the train displaying signals, where the movement can be properly protected by the register.

On roads where no intermediate registers are maintained, this rule has been altered to permit the dispatcher to display signals from an intermediate point, first protecting the movement by issuing a copy of the signal order to all trains affected. However, if a train is to be run to an intermediate point as a section of a schedule, they require that signals be displayed across the entire subdivision and the second section, which is to be discontinued, annulled by using Form K, from the point where it stops.

The old practice of entrusting the protection of such movements to the operator, or switch tender, or of a train having to leave a flagman, was never very satisfactory nor safe. It has been abandoned on most roads.

On single track, extra trains cannot be run without train orders. The dispatcher must authorize each movement; and he must protect that movement against any conflicting movement he may authorize later. On two or more tracks, we frequently find the authority for extra trains to move without train orders; their movement being with the current of traffic. Here is one rule showing how this is done:

D-97 (A). When authorized by special instructions in the time-table, trains moving with the current of traffic may run extra or work extra without train-order authority, except when a train order Form D-S is in effect, but must obtain a clearance before commencement of trip if at an open train-order office.—Southern Pacific.

RULES 99 and 100

Probably more wrecks have resulted from failure to comply with Rule 99 than from any other single cause. A number of factors are involved. To begin with, flagging is frequently done under more or less stress. This is the reason why a flagman's instructions should be in writing. It is not enough that they be instructed verbally.

A case is recalled where a flagman notified an engineman that his work train was coming to that station for him. Afterward this engineman received a clear block and proceeded, killing seven and injuring fifteen on the work train.

On another occasion No. 5 was delayed two hours at a station by No. 16 doubling in. The conductor instructed his brakeman to notify No. 5 to follow his engine back to D, where the rear part of No. 16 would wait, but the flagman became confused and as there was doubt in the mind of the engineman of No. 5 as to just what was expected of him, he decided to take the safe course.

Another case is recalled where the conductor of First 17 found it necessary to ride to the next station on Third 13 to protect his own train by flag. Upon arrival at the station in advance he failed to get off at the first switch to hold No. 18 and, becoming confused, instead of instructing No. 18 to wait, he merely identified himself as First 17.

Still another case is recalled where a conductor desired to have his flagman hold No. 20, but permit other trains to come to the point where he was waiting, without delay. The flagman became confused and held all opposing trains, thus tying up the road for the time being.

The majority of the roads adhere strictly to Code Rule 99; however, there are others where something of a departure is made in order to employ an audible stop signal. It will be remembered that under the Southern Pacific rules, the explosion of one torpedo is a signal to stop. Here is the S. P. flagging rule:

99. Conductors and engineers are responsible for the protection of their trains in both directions.

When a train stops under circumstances in which it may be overtaken by another train, the flagman must go back immediately with flagman's signals a sufficient distance to insure protection.

If recalled from a point less than one-fourth mile from rear of his train, he must, if safety to train requires, leave a lighted fusee, and, if conditions warrant, also place two torpedoes on the rail three rail-lengths apart.

If not recalled, one-fourth mile from rear of train he must place one torpedo on the rail; three-fourths mile from rear of train, or when recalled, if one-fourth mile or more from rear of train, he must place two torpedoes on the rail three rail-lengths apart. If conditions, such as curves, foggy, or stormy weather, or descending grade require, he must continue back a greater distance, placing two additional torpedoes.

When a train is seen or heard approaching when flagman is recalled, or before he has reached required flagging distance, he must immediately place one torpedo on the rail, and go toward the approaching train, displaying stop signals. Lighted fusees must be displayed when conditions warrant.

When flagman has reached the required flagging distance and has placed torpedoes as required, he may then return to the single torpedo where he must remain until relieved by another flagman or recalled. When recalled he may remove the single torpedo and return, leaving a lighted fusee when conditions warrant.

The front of the train must be protected in the same way, when necessary, by the brakeman; if not available, the fireman.

When circumstances require, train must be moved forward a sufficient distance to afford protection.

When a train is moving under circumstances in which it may be overtaken by another train, the flagman must take such action as may be necessary to insure protection. By night, or by day when the view is obscured, lighted fusees must be thrown off at proper intervals.

Before a train or engine fouls the main track in moving out of a siding or other track, flagman must, if necessary, go back with stop signals a sufficient distance to provide necessary protection.

When day signals cannot be plainly seen, owing to weather or other conditions, night signals must also be used.

99 (A). During station stops, flagman of a train carrying passengers will take position not less than thirty feet behind rear of train except when required to go farther to afford protection.

S-99 (B). When a flagman is sent with specific instructions affecting the superiority of an opposing train such instructions must be in writing on the prescribed form. When sent by train, he must ride on the engine and show the flagging order to the engineer who must let him off at the point at which opposing train is to be restricted.

Flagging instructions must be written in duplicate,

FLAGMAN'S HOLD ORDER

..................................*Station* ....195..

Conductor and Engineman ....

Train No. ....*Engine No.*.................

will flag from ....*to*.......................

Wait at ....*until it arrives.*

....

....

...........................*Conductor of*...........

When a flagman is sent with specific instructions affecting the superiority of a train, the instructions must be written in duplicate, one copy to be given to the flagman and the other to be retained by the conductor, who will send it to the Superintendent. When sent on a train, the flagman must ride on the engine and show the flagging order to the engineman who must let him off at the point where the opposing train is to be held.

one copy given to the flagman and the other retained by the conductor, or engineer if no conductor, until movement is completed, and then mailed to the superintendent. Flagging orders issued by conductor must be shown to his engineer.

An engineer must not carry a flagman against an opposing train, unless he presents his flagging instructions written on prescribed form. After reading the flagman's hold order, engineer must return it to the flagman.

Under the Standard Code Rule 99, nearly every detail as to flagging is left to the judgment of the flagman. In railroad operation, the flagman constitutes the highest authority for train movements. A train can go anywhere at any time under protection of flag. No one can countermand the authority of a flagman, except the conductor who sent him out.

It is not good judgment for a train to flag away from a train-order office or other point where the dispatcher can be contacted. The dispatcher is helpless in the face of a movement made under flag protection; and this often disrupts the entire subdivision.

If a flagman is sent out with instructions to hold all except first-class trains, no consideration should be given to the makeup of a train. All trains should be stopped; then those of the first class allowed to proceed. To assume that a train composed of passenger equipment is a first-class train is to take a chance. It may be a train of deadhead passenger equipment running extra or on a second-class schedule. Freight trains are sometimes run on passenger schedules.

The Code rule omits any reference to the distance a flagman shall go back in protecting the rear of his train. Conditions vary so widely that it is difficult to establish a safe distance, unless that distance be the maximum that could ever be required under the worst of conditions. There are, however, a few principles that should be adhered to by flagmen.

A flagman should not station himself on an obscured curve. He must always give consideration to the distance from which an approaching train can see his signals, continuing back to straight track where possible. On descending grades and in stormy weather, the distance should be increased sufficiently to permit any train to stop, after having been flagged.

Torpedoes and fusees should be used exactly as the rules prescribe on the road for which you are working. If torpedoes are placed against a rail joint, there is less danger of their being knocked from the rail without being exploded. They are usually placed two rail lengths apart, certainly not more than three. If stormy weather, it is sometimes best to duplicate the torpedoes by placing one on each rail, thus assuring that at least one of them will be exploded.

On roads where one torpedo is a stop signal, the flagman should never allow a train to pass him without first placing one torpedo on the rail, even though his stop signal may have been acknowledged.

When an engineman strikes torpedoes, he has no way of knowing when they were placed; however, a burning fusee tells him that a train has been there less than ten minutes before. A fusee dropped sufficiently in advance of the point where a stop is to be made affords protection during a short stop. If the train be delayed after making the stop, this gives the flagman protection while he is going back.

The enginemen, by dropping off a fusee when an unusual stop is to be made, can accomplish two things. He notifies the rear of the train of his intention to stop; and at the same time affords a certain amount of protection.

Once a flagman has gone back to protect, a train should not be backed without knowing positively that the flagman is out far enough to protect the movement.

The time elapsing between the time the flagman is recalled and the time the train is actually moving is critical. Where there is an obscured curve immediately to the rear, it is often good judgment to move the train ahead for a short distance.

The flagman must take necessary precautions when returning to his train, leaving a lighted fusee when necessary to protect his return to the train.

If a train is seen or heard approaching after flagman is recalled, he must not return but must continue in the direction of the train, displaying stop signals.

A flagman should never signal a train with both the white and red lanterns held in one hand. When so held together, the red light is not visible from a distance.

Should a train that has been flagged pass the flagman, he should not follow it but should remain out and make every reasonable effort to protect the rear of that train until relieved by its flagman.

Enginemen should whistle out flagmen whenever it is known their train requires flag protection. However, the flagman must not wait to be whistled out, or be sent out by the conductor. It is his duty to go at once without instructions from anyone.

Enginemen, when flagged, should stop and ascertain the reason therefor.

Arrangements should not be made between the crews of two trains for the following train to look out for the leading train at any point, thus eliminating the necessity for the leading train to flag. It is all right that the following train be told of an anticipated stop, but this does not relieve the leading train from the observance of Rule 99.

If an unattended red signal is found by a train, the

rules must be strictly observed, sending a flagman the full required distance, when necessary. It may have been placed there by an employe who discovered some defect in the track and had to protect in both directions as best he could.

There are certain places in CTC, as well as in other signal territory, where the rules of a road suspend flag protection, depending on the indication of signals to protect a train standing in these limits. Numerous collisions have resulted from this practice.

ENGINEMEN may as well know that, in every instance where a collision occurred because of inability to control the speed of the following train in accordance with signal indication, the I. C. C. reports have laid the blame right in the lap of the engineer—FAILURE TO OBEY SIGNAL INDICATION.

Obedience to signal indication becomes of the utmost importance when safeguards, such as Rule 99, are suspended.

Flagmen, and others, must keep in mind that a red fusee can be seen much further than a red lantern, especially during rain, snow or fog, and should make use of fusees when there is any question of the red light being seen in time.

When speed is materially reduced, fusees should be thrown off at proper intervals to afford ample protection. Flagman should watch carefully to see that the fusees continue to burn after they hit the ground.

Should a flagman be left behind, another member of the crew must assume his duties. If none is available, then the conductor must do the flagging.

RULES 101 to 108, inclusive, D-151 and D-152

There is no Standard Code Rule 100; however, the student will find a rule of that number in the Uniform

Code, used by the Missouri Pacific and several other roads. In recent years, there has been a tendency to include rules in revised books that are not to be found in the Standard Code. Some roads have shifted material from one number to another. As long as all essential material is retained, the over-all result is a set of rules which are safe, even if they vary from the Standard Code in many respects. The criterion by which a rule book is judged is simply this: Does it provide for safe, efficient transportation?

On the other hand, a person could not select one rule from each of a hundred or so books and come up with a good set of operating rules. Each book contains rules which dovetail together to produce the desired result. Therefore, in the final analysis, the student must stick to one set of rules—the rules of the road where he is employed. The material shown below in Rule 100 may be given elsewhere in your book of rules:

100. When an engine leaves a portion of its train on a main track under conditions which may make it difficult for the returning engineer to locate the standing portion, two torpedoes must be placed not less than 20 car lengths in advance of the head end of rear portion to serve as a warning. When conditions require, a trainman must protect returning portion.

When for any reason an engine leaves its train or part of its train on the main track, a sufficient number of hand brakes must be set, when necessary, to keep train from moving.

On heavy grades, when stopping on the main track or a siding, when cutting engine off a train or cars at stations to do work, or at stops of unusual length, with engine detached, a sufficient number of hand brakes must be set to hold the train or cars, and the air must be released.—Missouri Pacific.

Rule 101 is found considerably amended in the majority of rule books. Here is the entire Uniform Code rule:

101. Trains must be fully protected against any known condition, not covered by the rules, which interferes with their safe passage.

Conductors and engineers must inform themselves of conditions, and during and after excessive rains, heavy storms, fogs, or any condition which may restrict visibility or affect condition of track, must restrict speed of their train to insure absolute safety, and if in doubt of being able to proceed safely, train must be placed on siding until it is safe to proceed.

When storms, fogs or other conditions obscure track or signals, from points where they are plainly seen under normal conditions, speed must be restricted to insure seeing and complying with indications of any and all signals, regardless of loss of time.

101 (a). Trainmen must know by speed of train, grade or caboose air gauge, that train is being handled safely and under control, and, when necessary, take immediate action to get train under safe control.

If any crew member of a train has reason to believe the train has passed over any dangerous defect, the train must be stopped and protection afforded.

101 (b). Whenever a part of any bridge has been damaged, no train or engine shall pass over the structure until it is known that it is safe for traffic.—Missouri Pacific.

Referring again to Rule 100 of the Uniform Code, this material is given in the Santa Fe rule book under 102 (A), which reads:

102 (A). When it becomes necessary to leave a portion of

a train on a main track between stations, two torpedoes must be placed on the rail on the engineman's side about 1000 feet in advance of the forward end of the detached portion. In addition, by night, a red light must be placed in a conspicuous position on the forward end of the detached portion.

Herein is reflected the difference of opinion between carriers as to the proper method of doing things, as well as what they deem the best manner of wording their rules.

Old Code Rule 102 relative to a broken-in-two condition on single track has been eliminated from the latest edition, as well as signals 12 (e) and 14 (f), the "train parted" signals. This is on the assumption that present-day air brakes do not permit of such a condition.

On the other hand, some roads still retain these in view of increased air-pump capacities, which sometimes forestall emergency braking action. Therefore, the discussion relative to a broken-in-two condition is retained.

The rear portion of a train should not be left on a siding at a station depending entirely upon markers and the absence of the front portion to protect the return movement, when necessary to double to some siding in advance. A competent man should be left at the station to hold or otherwise arrange with following trains.

Should a freight crew notice a caboose displaying markers at a station by night, no trainmen about and no engine attached, it would be well to stop and investigate, even though the switches are lined for the main track.

Should any train find only the rear part of a train at a station and the switch at the front of such train lined for the siding, then an explanation is necessary before proceeding.

A good many roads have thought it essential to enlarge upon Code Rule 103, to provide for more precise instructions as to switching movements, necessary precautions, etc.

The same is true of Rule 104, especially on roads where spring switches, dual control switches and electrically locked switches are in service.

Shown below is Rule 104 in its entirety, taken from Uniform Code of Operating Rules, subscribed to by the Missouri Pacific and other roads:

> NOTE: Rule 104 (not including Rules 104 (a) to 104 (f), inclusive) applies only to hand-operated switches. When spring or remote-control switches are operated by hand, they are then hand-operated switches and rules governing hand-operated switches apply.

104. **HAND-OPERATED SWITCHES**

(1) Main-track switches must be lined and locked for main track when not in use. Other than main-track switches, equipped with switch locks, must be lined and locked for normal position when not in use. ●

The following other than main-track switches must be kept lined in normal position, except while movement through them is being made:

(a) Crossover switches.

(b) Switches connecting other tracks with a siding.

(2) Except as prescribed by Rule 402, main-track switches must not be left open after movement through them is completed, unless attended by a member of the crew.

(3) A main-track switch must not be left open for a following train or engine, unless in charge of a member of the crew of such train or engine, or an assigned switchtender.

(4) When practicable, the engineer must see that switches near the engine are proper lined and must require other members of crew on engine to observe position of such switches.

(5) A train or engine must not foul a MAIN TRACK OR OTHER TRACK until switches connected with the movement are properly lined.

When waiting to cross from one track to another and during the approach or passage of a train or engine on tracks involved, all switches connected with the movement must be secured in the normal position.

Main-track switches must not be restored to normal position until the movement is completed or clear of the MAIN TRACK involved.

(6) Where trains or engines are required to be reported clear of the main track, such report must not be made until switch has ben secured in its normal position.

Main-track switches must not be restored to normal position until the movement is completed or clear of the MAIN TRACK involved.

(6) Where trains or engines are required to be reported clear of the main track, such report must not be made until switch has been secured in its normal position.

(7) After restoring a main-track switch to normal position, employe must test the lock to know that it is secured and see that switch points fit properly. Defective or missing main-track switch locks must be replaced immediately or switch securely spiked for main-track movement.

(8) Derails must be set to derail, and except pipe connected derails, must be locked (if equipped with

locks) in that position, unless lined to permit movements.

(9) After lining a main-track switch for a train, the employe attending the switch must go to the opposite side of main track, when practicable, and not return to the operating switch stand until the movement has been completed.

When not practicable to go to opposite side of track, the employe will stand at least 20 feet from operating switch stand.

(10) Employes alighting from a moving train to restore main-track switch to normal position, must, when practicable, get off the rear end of rear car, on opposite side of train from the operating switch stand, and must not cross over to switch stand until it is in clear.

(11) When a train or engine is clear of main track to meet or be passed by a train, employes must not unlock, nor take a position in the vicinity of any main-track switch. They must not go beyond the clearance point for the purpose of attending the switch to be used, and must remain at least 150 feet from the switch while the expected train is approaching or passing the switch.

(12) Employes throwing switches must see that both points have moved and fit in proper position, that lever is properly secured, and when operating lever is equipped with latch, they must not step on latch, except when throwing switch.

(13) Hand-operated switches must not be run through. When a hand-operated switch is run through, it is unsafe, must be protected, and must be spiked unless section foreman takes charge at once. If an engine or a car partially runs through

such a switch, the entire movement must be continued.

(14) Scale-track switches must be lined for dead rails when scales are not in use.

(15) At main-track switches in ABS territory, where view is not clear for at least one mile in each direction, trainmen and yardmen will operate switch and WAIT 3 MINUTES at the switch before giving signal for train or engine movement to main track, EXCEPT:

(a) Where switch is equipped with an electric lock.

(b) Where block signals GOVERNING MOVEMENT TO MAIN TRACK indicate proceed, or block indicator indicates clear.

(c) Where signals on MAIN TRACK indicate proceed in direction of restricted view, with clear view in opposite direction.

(d) At meeting points where switch is operated before the train met has passed its next signal.

(e) When entering the main track between signals to hostle engine or switch train standing between such signals.

The 3-MINUTE WAIT does not relieve employes from protecting the movement.

(16) Main-track switch targets will show RED when switch is lined for movement to or from main track.

104 (a). **SPRING SWITCHES**

Spring switches are properly designated and facing point movement over them protected by signals.

When signal in facing point movement displays "Stop" or "Stop, Then Proceed at Low Speed" indication:

(1) Test switch by throwing over and back by hand, examine switch points to see that they fit properly, and that switch is lined for route to be used.

(2) Train or engine may then proceed as prescribed by Rule 350 or Rule 351 (Automatic Block Signal Rules).

A train or engine trailing through and stopping on a spring switch must not make reverse movement, nor take slack, while any part of train or engine is on switch point until switch has been thrown by hand.

Where there is no signal protecting trailing movement to a main track through a spring switch, or when signal governing movement to main track indicates "Stop" or "Stop, Then Proceed at Low Speed," or when block indicator shows block occupied, the main track must not be fouled until it is seen that track is clear and protection afforded against following trains or engines on that main track.

104 (b). **DUAL CONTROL SWITCHES**

When authority has been granted and track and time limits secured from control operator in territory where rules governing movement of trains and engines by block signals are in effect, and when permission has been obtained from control operator in other territory to operate a dual control switch by hand, switch may be operated as follows:

(1) Unlock switch lock.

(2) Operate dual control selector lever marked "Power" or "Motor" to position marked "Hand."

(3) Operate hand lever back and forth until switch points are seen to move with movement of

lever, then line switch in position for route to be used.

(4) After movements over switch have been completed, restore switch by hand to normal position, then lock dual control selector lever in position marked "Power" or "Motor" and notify control operator.

The authority granted by control operator to operate a dual control switch by hand does not authorize any part of the train or engine to move beyond the track limits, and does not authorize movement over such switch, except on hand signals from trainman stationed near the switch.

Before making movements over the switch, trainman must notify engineer when dual control selector lever is in "Hand" position, and engineer will then be governed by hand signals, as block signals indicate "Stop" while selector lever is in "Hand" position.

When dual control selector lever is restored to "Power" or "Motor" position, engineer must be notified and he will then be governed by block-signal indication.

Note: Operation by hand of remote-control switches other than dual control switches will be authorized by control operator, and where instructions are posted near switch, they will govern.

104 (c). **EXAMINATION OF SWITCHES**

Before proceeding from a stop-indication over a remote-control switch under provisions of Rule 350, trainman must examine switch, see that switch points fit properly, and he must remain at switch until leading wheels pass over switch.

104 (d). **ELECTRICALLY-LOCKED HAND-OPERATED SWITCHES**

The operation of electrically-locked hand-operated switch will be governed by general order, special instructions in time-table or pamphlet form, or instructions posted near the switch.

104 (e). **REPORTING DEFECTIVE SWITCHES**

Immediate report must be made to train dispatcher, and, when practicable, track foreman notified, when switches or switch locks are damaged or found defective, or when hand-operated switches are run through. When necessary to spike a spring switch, or when any condition exists affecting a switch which imperils movement of trains or engines, switch must be protected.

104 (f). **USING SAND AND WATER**

Sand must not be used or water allowed to run from engine appliances over spring, remote-control or interlocked switches.

Rule 105 is self-explanatory. Actually the authority to use a siding of assigned direction in the reverse direction comes from the dispatcher, regardless of whose signature is shown.

Rule 106 places the responsibility for the safety of the train equally with the conductor and engineman.

Rule 107 appears in various forms in different rule books. The intent of the rule is to keep a train from running between a passenger train which is discharging passengers and the station where the passengers are going, which might endanger the lives of passengers unless some safeguard is provided.

Rule 108 should be applied to every phase of railroad

operation, since it is always better to be safe than sorry. Doubt or uncertainty may arise out of practically every phase of railroad operation and sufficient precautions should be taken to avoid hazard.

Rule D-151 usually provides that trains must keep to the right; however, there is at least one road operating in this country which requires its trains to keep to the left.

Rule D-152 must be understood to mean that a train crossing over to or obstructing another main track must provide flag protection in both directions, not just in the direction of the current of traffic. The dispatcher may be running some train on that track against the current of traffic.

RULES 201 to 206, inclusive

Rule 201 usually provides that train orders will be issued over the signature of the chief train dispatcher; however, they may be issued over the signature of any designated official from the train dispatcher to the superintendent.

The provision that they must be issued in the prescribed forms when applicable gives the dispatcher enough leeway to word his orders in the clearest possible manner. He is required, of course, to use prescribed forms where these can be made to cover the situation.

Rule 202 was written to put a stop to the use of "single orders," or the transference of part of one order to an order of a different number by the dispatcher. Orders must be given to all trains affected under the same number, and all copies must be identical in their contents. No system of handling train orders has the maximum element of safety if this rule is not strictly adhered to.

Train orders, with the exception of slow orders, are numbered consecutively each day, beginning at midnight. Some roads permit the use of an entirely different series of numbers for slow orders and these are not revised daily, continuing in effect until annulled.

Where more than one set of dispatchers work in a single office, each set begins at midnight with a different series of numbers, i.e., 1, 201, 401, etc. The dispatcher must carefully avoid duplicating order numbers; and no numbers should be skipped, such as "13."

Rule 204 will be found in varied form in various rule books. Below is the entire rule, as shown in the Santa Fe rule book:

204. Train orders must be addressed to those who are to execute them, naming the place at which each is to receive his copy. Those for a train must be addressed to the conductor and engineman, and also to anyone who acts as its pilot.

Not more than one address will be shown on a line in train orders or in dispatcher's order book.

A copy for each employe addressed must be supplied by the operator. When more than one engine is attached to a train, each engineman must be supplied with copies of all orders affecting the movement of the train.

An additional copy of all train orders and clearance cards must be furnished for delivery to rear brakeman.

Orders addressed to operators restricting the movement of trains must be respected by conductors and enginemen the same as if addressed to them.

The usual procedure in recording train orders under the provision of Rule 205 is about as follows:

Where dispatching is done by telephone, the dispatcher writes the order in the train order book as he transmits it, underscoring each word and figure as it is repeated by each operator, recording the times repeated and completed, and the names of those who sign for the order.

Where dispatching is by telegraph, the dispatcher sends the order, copying it in the order book as it is repeated by the first operator. Following repetitions are then underscored in the same manner as described above and the same information recorded.

The majority of roads in this country still use two order forms, namely, "19" and "31." The use of form "19" has been liberalized on most roads, the "31" being used only when necessary to restrict a train under unusual circumstances where there exists doubt as to the delivery of the "19" form.

The Southern Pacific and Santa Fe have used a single form for several years, it being designated as a "Train Order."

The rules outlining the proper transmission, repetition and handling of train orders are quite clear and should be studied carefully by those entrusted with this responsible duty. It is most exacting work, where a single mistake may cause a wreck.

Some roads require that time be spelled out in train orders, then duplicated in figures. Other roads require that only figures be written in train orders, and in the dispatcher's order book; however, in transmitting and repeating orders, time is spelled in the usual manner. For instance, the dispatcher says "nine one, n-i-n-e o-n-e, 9-0-1 A M"; but he only writes the figures 901 AM in the order book. The operator writes only the figures in the train order but, in repeating, follows the same procedure as the dispatcher used in sending.

It is the generally accepted practice, where two or more engines are coupled, to use only the leading engine number in train orders. If a helper engine is coupled on the head end for a short help, then the number of the through engine is used. As a rule, helper engines cut back in a train are not mentioned in train orders, although there is at least one road which requires that the numbers of all engines in a train be shown in train orders; for instance: "No. 441 Eng 314 and Engs 315 and 316 assisting."

The majority of rule books will be found to vary in some respect from the Standard Code Rule 206. For instance, the multiple-unit diesel-electric freight engine, with each unit numbered, has presented a small complication. For comparative purposes, Rule 206 from the Missouri Pacific book is presented in its entirety:

206. **DESIGNATION OF TRAINS**

In train orders regular trains will be designated by number as "No 10" and sections as "Second 10," adding engine numbers.

Extra trains, except work extras, will be designated by engine numbers, and the direction as "Extra 798, 'North' or 'South'."

Work extras will be designated by engine numbers, as "Work Extra 798."

For the movement of an engine of another company, the initials as well as the engine number will be used.

The identifying numbers on the operating control units of diesel engines must be displayed and the identifying numbers on the nonoperating control units must be concealed while in road service.

When two or more engines are coupled, the number of the leading engine only will be used in train orders, except the numbers of helper engines used over a portion of the subdivision need not be shown.

206 (a). **TRANSMITTING AND REPEATING**

Even hours as "10 00 A M" must not be used in stating time in train orders.

In transmitting train orders, time will be stated in figures only.

In transmitting or repeating train orders by telephone, the numbers of trains, engines, orders and other numerals, except time, must be pronounced as whole numbers, then each figure pronounced separately, thus: One Thirty-Five, One-Three-Five; except where there is but one figure, it must first be pronounced, then spelled, thus: One, O-N-E. Names of stations and direction of extra trains must be pronounced, and then spelled letter by letter, thus: Aurora, A-u-r-o-r-a; North, N-o-r-t-h. Time in train orders must be pronounced as a whole number and then each figure spelled separately, thus: One Fifty, O-n-e F-i-v-e N-a-u-g-h-t.

The letters duplicating names of stations and numerals will not be written in the order book or upon train orders.

When train orders are transmitted by telegraph, the train dispatcher must write the order from the first repetition and underscore each word and figure at the time of each succeeding repetition. When transmitted by telephone, he must write the order as he transmits it and underscore each word and figure at the time of each repetition.

When two extras are mentioned in train orders the word "two" shall be used, for example:

"Meet two Extras 798 and 799 North."

206 (b). A train order may be transmitted to conductor or engineer, in which case such employe copying order will be governed by rules applicable to operators governing repetition and completion of train orders.

If a restricting order is sent in this manner, signatures of conductor and engineer of train restricted must be received by train dispatcher before "Complete" is given to order for the other train.

206 (c). **RELAYING**

To relay a train order, it must be transmitted in the usual manner to the relaying operator, who must transmit the order to office addressed.

The operator at the office addressed must repeat the order to the relaying operator who must underscore upon his office copy each word and figure as repeated by the office addressed.

The relaying operator must then repeat the order to the train dispatcher, by whom "time complete" will be given to the relaying operator to be transmitted to the office addressed.

RULES 207 to 214, inclusive

On roads having but one order form, Rule 207 is necessarily altered. Here is the Santa Fe rule:

207. To transmit a train order, the direction must be given to each office addressed, and the number of copies stated thus: "West, copy 4," or "East, copy 5."

The requirements of various roads with regard to a restricting order issued to a train at the point of restriction will not always be found under Rule 208, as in the case of the Code rule. In fact, the latest rule books show a definite tendency to shift material from rules of one number to those of another.

However, the principle may be set forth. When a train approaches a station unaware that an order is placed at that station which restricts its further movement, that train will come up the main track. Under certain conditions, as where the train order office is located near the end of the siding where the train must stop, precautions must be taken to prevent the restricted train from overrunning the point of restriction and colliding with another train moving to the station under the provisions of the order.

The student must examine his own rule book carefully so as to be fully informed as to what action must be taken under those circumstances. The majority of roads place a considerable responsibility with the operator, it being required that they display stop signals in advance of the office, where safety requires such action. Regardless of the order form used, the train must be brought to a full stop before delivery of the order.

Rule 209 is fairly uniform in the various books with respect to copying orders. Certain procedures are outlined where errors are made in orders. Under no circumstances should an operator copy an order over, or make any kind of alterations after it has been made complete, without first repeating the new copies to the dispatcher. If the provisions of this rule are adhered to, a minimum number of errors will be found in the operators' work.

Some difference will be found in the rules of roads using a "train order," as compared to those roads using two forms, the "19" and "31."

There are still some roads in this country that require every restricting order to be on Form 31. On others, while still retaining the use of Form 31, said use is restricted to cases where the dispatcher must know the order is effective to a superior train before its delivery to an inferior train, for instance:

(1) When restricting a train by an order sent to it at a point not a train-order office, or one at which the office is closed;
(2) When restricting a train which has been cleared, or the engine of which has passed the train-order signal in proceed position;
(3) When restricting a work extra;
(4) When any condition makes it mandatory for the dispatcher to know a train is properly held before use of the order by an inferior train may be authorized.

Roads employing only one order form have rules which, in extreme cases, require the taking of signatures, either on the order itself or on a clearance which has space provided for this purpose.

There are two methods employed for the delivery of train orders. One is where the position of the train order signal is normally at "proceed"; the other is where the normal position of the signal is at "stop."

Where the signal is normally clear and the dispatcher tells the operator, "19 West copy 3," the operator must immediately put his westward signal at "Stop" and give the response "SD West." He may not again clear the signal until after the delivery of the order. If necessary, other trains wil be given a clearance on the signal.

An order addressed to the operator, fixing a meet at that station, requires both train order signals to be displayed at "Stop."

Copies of such orders, "middle orders," must be delivered to all trains affected until all have arrived from one direction. It makes no difference which direction.

Where any order is addressed to a train, the order board must not be cleared for any train in that direction until after delivery has been made, the schedule has been annulled, or the train has become more than twelve hours late on its schedule, or the order is annulled.

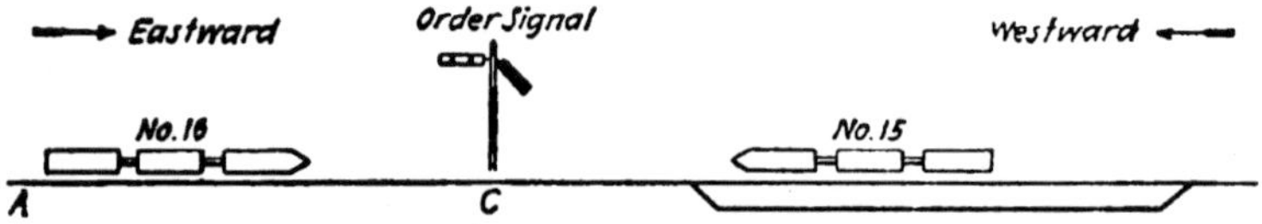

ORDER NO. 1

No. 15 get this order hold main track
and meet No. 16 at C instead of D.

Considering the above chart, it will be realized that issuing Order No. 1 constitutes something of a hazard.

No. 15, of course, should not pass the west siding switch before it knows why the train-order signal is at "stop."

On the other hand, No. 16 should take precautions, sending a flagman well beyond the west switch to protect its movement into the siding.

It will be noted that Code Rule 210 requires that the conductor personally deliver a "31" order to the engineer; and that the engineer read the order to the conductor. Both of these requirements have been dropped by some roads.

In addition, some roads have rules forbidding any member of the crew to read an order to, or tell any member of the crew the contents of an order. If each member of the crew reads the order for himself, they will not all read the order wrong; whereas, if one member of the crew misreads an order and tells the rest of the crew the contents before they have read it, they also are then likely to misread their instructions.

A work extra, for instance, can never be said to be properly held for an order until the dispatcher has some form of acknowledgment from the crew. And an operator must never tell the dispatcher that he has the work extra for an order until he at least has the conductor in the office and has his assurance that the engineer can be had. Even then, the dispatcher cannot definitely know he has the work extra, or complete the order to an inferior train, until after he has the signatures, or the order has been delivered.

A "19" order is just as effective as a "31" order, so long as it can definitely be placed so there is no hazard of non-delivery.

A dispatcher may instruct an operator to destroy an order which has not been repeated, or the X-response sent. After repeated, or the X-response given, it must be annulled if not to be completed.

The provisions of Code Rule 211 should be carefully studied in the individual rule book. Some roads require that clearances be checked with dispatcher and made OK by him; however, a few roads place sole responsibility for delivery of train orders with the operator.

The majority of roads still provide for an operator to X an order, even on roads still using a "train order." However, at least one of the newer rule books makes no provisions for X-ing an order, as provided by Code Rule 212.

Before a train dispatcher can confer right on one train, he must first take away the right of another train, or trains, to proceed. Rule 213 sets forth the fact that the dispatcher must make the order effective to the train restricted before making it complete to the inferior train. This is only common sense. If a dispatcher completed an order for an inferior train and cleared it without first restricting the superior train, communications might fail and he would be unable to make the order effective to the superior train, even though he had ample time in which to do this. A dispatcher must proceed on the assumption that the wires may fail at any moment, or that he may drop dead at any moment; to say nothing of a lapse of memory, which is a human trait.

Once a train order has been X'd, or repeated, it becomes a holding order for the train or trains addressed, even though not completed. If the line fails before the order is X'd, or repeated, it is treated as if the order had not been sent; that is, as if the order did not exist. If any part of an order has been repeated, the operator must consider it a holding order, even though it may be apparent that the dispatcher never heard the repetition because of line trouble.

There is no Code Rule 215. Various books do contain rules by that number. Here is one from the Missouri Pacific's book:

215. CHECKING ORDERS AND CLEARANCES BY CREW—Conductors, engineers, firemen and trainmen must read train orders and clearances, check the correctness thereof, see that the orders and clearances are addressed to their train, and that the order numbers shown on clearance correspond with the numbers of the orders received. Engineers must show train orders and clearances to firemen and to the trainmen on the engine, using the extra copy furnished engineer for this purpose. Conductors, when practicable, will show train orders and clearances to trainmen, except rear trainmen, who must be provided with a copy.

Members of the crew must, when practicable, check with each other, and have a definite and proper understanding of the requirements of train orders and clearances.

If necessary, firemen and trainmen must ask for train orders and clearances.

Any misunderstanding of a train order must be immediately corrected, even if necessary to annul such order and issue another.

The majority of rule books still contain Rule 216, setting forth the requirements where train orders are delivered by a train dispatcher. Some roads, however, vary these requirements, and their rule may be found under another number.

Code Rule 217 may be found considerably altered, especially on roads which use the "train order" instead of the Forms 19 and 31. The Santa Fe, for instance, does not permit a restricting order to be sent to a train at a point not an open train-order office, nor to one where the office is closed.

The rule appearing under Code No. 218 will be found in all rule books but not necessarily under that number.

It is probably one of the most misunderstood rules in the book. If an order is addressed to No. 22 and there turns out to be four sections of No. 22, then each section must have a copy of the order. Should the order state "No. 22 meet No. 21 at D," this is the equivalent of saying: "First Second Third and Fourth 22 meet No. 21 at D."

If the dispatcher should later issue an order "No. 21 meet First 22 at G instead of D," this would leave No. 21 still holding a meet with the last three sections of No. 22 at D, the meet having been superseded only with the first section.

Code Rule No. 219 will often be found altered in various rule books, especially the roads which use a single order form, to provide safety under their system of handling orders.

Code Rule 220 will be found with little variation, and is usually under that same number. It sets forth the specific life of each train order; that is, how long each order remains in effect after having been made effective.

Shown below are rules found supplementary to Code Rule 220.

Some roads rules forbid the operator to file orders that may become dead for any reason, such as a train becoming more than twelve hours late, or being annulled. The dispatcher must annul such orders; which gives a double check and assures that no live order will be filed by the operator.

The C. & N. W. rule places more stringent requirements on the crews of a train when one crew relieves another. The majority of roads do not require a written transfer.

The S. P. rule makes it possible for a crew to retain its slow orders, no matter how many times their identity may be changed during a tour of duty. The second paragraph sets up a condition not frequently found. If, for instance, a train dies on its schedule and has to be run extra, it may have some orders in its possession of a helping nature; that is, time on a first-class train, for instance. In order to save the dispatcher time, when creating the train as an extra, he may add onto their running order "Respect Order No. 155," instead of having to send Order 155 to the train when they have it in their possession, but under the rules cannot use it.

220 (a). A list of orders and instructions transferred from one conductor or engineman to another as required by Rule 220, must be made in duplicate in writing and checked, each retaining a copy.—Chicago & North Western.

220 (a). Train orders relating to track conditions, unless annulled, must be respected by conductor and engineer on all trips made during the tour of duty on which such orders are received.

A train order that does not restrict the train addressed may be made applicable to an additional trip by adding to the order creating the train for the additional trip, the words: "Respect Order No. ....—Southern Pacific.

The order system where the normal position of the train order signal is at "stop" is probably best set forth by the Southern Pacific rule, which reads:

221. A train-order signal must be provided at each train-order office, except at stations where all trains are required to obtain a clearance.

The train-order operator must keep the signal in stop position while on duty, except when he clears it for an immediately approaching train.

When a train order is to be delivered to an approaching train, or orders are held for any other train in the same direction, the operator must not clear the signal. If the operator does not hold an order for the train, or for any other train in the same direction, unless otherwise directed, he must clear the signal after the train has reached a point from which the signal can be plainly seen. After having been cleared for a train the signal must be restored to stop position as soon as practicable after the rear of train has passed.

When the signal indicates "stop" the train must not leave the station without a clearance; and, until the engineer has ascertained that he is not to receive an order which restricts his train at that point, it must not pass the fouling point of the switch at which an opposing train may enter the siding. If operator has no order restricting the superiority of any train in the same direction at that point, he may lower train-order signal arm twice and immediately return it to stop position as an indication that the train is to receive orders which do not restrict it at that station, or is to receive a clearance without orders; but outside of block system limits, except when a clearance only is to be delivered, the operator must obtain permission from the train dispatcher before thus operating train-order signal for an approaching train. The operator will then make delivery without causing the train to stop. After receipt of clearance, and orders if there be any, train may proceed without stopping, unless restricted by time-table, or by train orders previously received.

When it is apparent that orders are to be received without stopping, the train-order office must be passed at a speed not exceeding thirty miles per hour.

If the signal is first seen in proceed position, or if after changing to "proceed" it is changed to "stop" before the train passes it, the train must not proceed without a clearance.

Outside of block system limits, operators must space trains ten minutes apart when moving in the same direction, using the train-order signal and clearance for that purpose. When necessary to space trains in this manner the time the following train will be permitted to proceed must be shown on the clearance in the space provided for that purpose, and conductor and engineer must respect that time.

Train-order office hours may be shown in the time-table. When not so shown, and the train-order signals for both directions indicate "proceed," and in addition a green light is displayed by night, it will indicate that the train-order office is closed.

When light is not displayed in a train-order signal at night, where under the rules one should be displayed, unless a clearance is received or both arms of the signal can be plainly seen in proceed position, train must stop and determine indication. If either arm be in stop position, a clearance must be obtained when train-order operator is on duty; if no operator is on duty train dispatcher must be consulted before proceeding.

221 (a). When a clearance is issued to a train the numbers of all train orders, if any, addressed to the train must be shown thereon, and conductor and engineer must know before leaving that the numbers shown correspond with the orders received, and that all information required on the clearance is properly shown thereon.

Before delivering train orders, operators must carefully read the addresses of each train order. fill out the clearance, entering thereon, without alteration or erasure, numbers of all train orders for the train, transmit the address and order numbers to the train dispatcher, who will check the correctness thereof against his record in the train-order book; give the OK, time and initials of the chief train dispatcher, and make proper record thereof, and the operator must enter this information on clearance, after which, clearance with orders will be delivered.

Space for time, following space for operator's name on clearance, will not be used when clearance bears OK, time and initials of chief train dispatcher.

If necessary to restrict a train as prescribed by Rule 211, after delivery of clearance to conductor or engineer, or both, all copies of the clearance must be destroyed by the operator and a new one issued showing thereon all orders for delivery to the train at that station.

If, after a clearance is delivered to conductor, or engineer, or both, orders are issued for a train which do not restrict its superiority, the train dispatcher, without requiring the operator to destroy the clearance first issued, may authorize issuance of a second clearance which must be endorsed "Second" preceding the work "clearance" on top margin. All orders for delivery to the train at that station must be shown on second clearance.

For the purpose of comparison, shown below is the same rule from the Uniform Code book, which prescribes that the normal position of the train-order signal is clear.

221. TRAIN ORDER SIGNALS.—Unless otherwise provided, a fixed signal must be used at each train-order office, which shall display indication per Rule 232 (Stop Train Order Signal) when clearance is to be delivered, or when blocking trains per Rule 91 (10-minute block for following trains).

A train must not pass the train-order signal displaying indication per Rule 232 without a clearance, except to do station work or to enter siding, protecting when required as prescribed by Rule 99, but must receive clearance before leaving.

When engine crew can see that clearance, or clearance and orders are in position for delivery, it will indicate such delivery will be effected.

When an operator at station equipped with a train-order signal receives the signal "Copy" followed by the direction, he must immediately display train-order signal indication per Rule 232 for the direction indicated, and then reply "SD," adding the direction, and until the orders have been delivered or annulled, the signal must not be restored to "Proceed" indication.

Each train order must be delivered to the train addressed, unless the order is annulled by the train dispatcher.

221 (a). "CALLING-ON" INDICATION.—To advance a train beyond a station where it is restricted for a train, train dispatcher will first issue necessary order to advance train, and then issue an order, Form N, to operator to display "calling-on" indication of train-order signal, if train to be advanced is first from its direction.

Operators must not display "calling-on" indication of train-order signal, except when authorized by train order, Form N.

A train approaching a station when "calling-on" indication is displayed will acknowledge as prescribed by Rule 14 (b) (two long blasts of engine whistle), and use main track at restricted speed to train-order signal.

When engineer sounds whistle signal as prescribed

by Rule 14 (b), in answer to "calling-on" indication, operator will again display train-order signal indication per Rule 232. Train orders and clearances must be delivered with signal indication per Rule 232, and train must not leave station while "calling-on" indication is displayed.

If a train is delayed between the time of accepting the "calling-on" indication and arrival at the train-order signal, train must be protected in both directions, as prescribed by Rule 99.

The "calling-on" indication of train-order signal may be repeated at designated points, at or near switch where train would enter siding, by a yellow light identified by the letters "CO."

The Missouri Pacific is not the only road using the "calling-on" signal to bring a train up to the train-order signal on the main track when it would normally be required to take siding at the station on orders previously received. A yellow signal is generally used for this purpose, either in connection with a third position of the train-order signal, or otherwise.

To show how Code Rule 222 is filled in, the Southern Pacific rule is reproduced below. Again, this rule differs considerably from the rules of other roads, emphasizing the need for each reader to check his own book for exact requirements.

222. Operators must promptly record, and report to the train dispatcher, the arrival and departure of all trains, reporting "no signals" or "green signals," as the case may be. If report of signals displayed is incorrect, the train dispatcher must immediately notify all trains affected until he has ascertained that the signals are properly displayed, and should "no signals" be registered when "green signals" should

have been registered, he must require immediate correction in train register.

222 (A). Operators in relieving each other must make a transfer on prescribed form, of outstanding train orders and undelivered messages addressed to, or in care of, trains; also show on the transfer, except at train-register stations where all trains register, the numbers of all overdue trains and whether trains then in the yard or at station have or have not been cleared. The operator assuming duty must not handle train-order signal, nor deliver train orders, until transfer has been made and signed.

If an operator is permitted to close his office, leaving track orders, annulments of schedules, or undelivered messages addressed to, or in care of, trains, for another operator coming on duty at a later time, he must list them in transfer book and operator coming on duty later will sign it.—Southern Pacific.

In addition to those abbreviations authorized by Code Rule 223, many others have been adopted by various carriers. Here are some of them:

ABS	—for Automatic Block Signals
AABS	—for Absolute Automatic Block Signals
CTC	—for Centralized Traffic Control
ATC	—for Automatic Train Control
ATS	—for Automatic Train Stop
Dist	—for District
Div	—for Division
Subdiv	—for Subdivision
MP	—for Mile Post
MPH	—for Miles Per Hour
O K	—for Correct
S D	—for Stop Displayed

Int	—for Interlocking Station
MB	—for Manual Block Station
Condr	—for Conductor
Engm	—for Engineman
9	—to Clear the line for Train Orders, and for Operators to ask for Train Orders.

Others, no doubt will be found in various rule books.

TRAIN ORDER SIGNAL

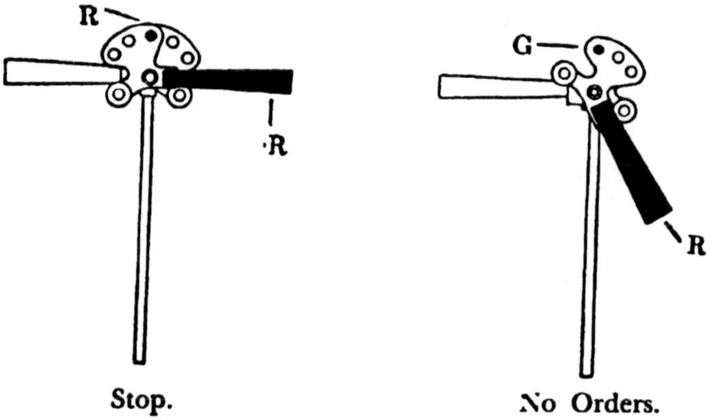

The arm to the right, as seen from an approaching train, is the one which governs. When the arm is extended horizontally, or in addition a red light is displayed, it indicates "stop." When the arm is inclined downward, or in addition a green light is displayed, it indicates "proceed." When both arms are inclined downward by day, and in addition a green light is displayed by night, it indicates a closed office.

When a train has been cleared, or its engine has passed the order signal in clear position, extreme caution must be exercised by all concerned should it become necessary to issue an order restricting the train. The rules of the various carriers outline safe procedure and they should be strictly followed.

Should there be a complete failure of communications, the rules of the majority of roads permit the operator to deliver to a train such orders as he may hold, together with a clearance endorsed "Wire Failure."

A clearance card, dated, timed and OK'd before midnight may be accepted by a train after midnight. Often this occurs when a train is expected to leave before midnight but is delayed.

CLEARANCE CARD

Form A

........................*Station* ....*19* ..

Conductor and Engineman..

Orders for you are..

....

(If no orders, endorse "NONE" on first line of space provided for order numbers.)

O. K. at....................*M*................*Chief Train Dispatcher.*

Do not leave before..*M*
(Fill in this line only when necessary to comply with spacing requirements.)

........................*Operator Time*........*M.*

Conductor and each engineman must have a copy and see that their train is designated correctly, also that the numbers of all train orders received correspond with numbers inserted above.

Clearance card must be filled out by the operator before repeating train and order numbers to the train dispatcher.

After receiving "O. K." from the train dispatcher no additions must be made, but a new clearance card issued.

Operators must retain a carbon copy.

Where the train-order signal and clearance is used to space trains, the requirements vary as to whether the train should be brought to a stop before delivery of a clearance, or whether the clearance may be handed up, allowing the train to pull down to the switch to wait for time.

When a "31" order is sent restricting a train at a blind siding, six copies are usually required: Two for the crew in whose care the order is sent, two for the train at the blind siding, one upon which to take the signatures of the conductor and engineer of the restricted train, and an office copy. The order with the signatures must be left at the first available point of communication by the train in whose care the order is sent.

Chapter IX

FORMS OF TRAIN ORDERS

GOVERNING PRINCIPLES

Note: Forms with a prefix "S" are for single track, those with a prefix "D" are for two or more tracks; those without prefixes "S" or "D" are for single or two or more tracks. The prefixes "S" and "D" to be printed in italics. The words and figures in italics in the forms are examples indicating the manner in which the orders are to be filled out.

In making a study of the application of train orders, certain governing principles should be kept in mind.

First. Form S-A is positive in meaning. There is no such thing as a meet order on two or more tracks. A train has not been met until its markers arrive, or someone accounts for their absence. If the expected train does not arrive, the train waiting at a meeting point must wait until:

(a) The meet is superseded or annulled;

(b) The schedule of the train becomes more than 12 hours late;

(c) Annullment of the train to be met is received; or,

(d) A new time-table takes effect, showing there is no such opposing train.

Second. A train order is not in effect until after it has been X'd or repeated. Once effective, it continues so until fulfilled, superseded or annulled.

Third. There is only one way to supersede an order, and that is by the use of the words "instead of." A later order must not be confused with a superseding one, received earlier, with which it conflicts, unless it contains the words "instead of."

Fourth. When an order has been superseded, it becomes void, the same as when annulled. The order cannot be restored by annulling the superseding order.

A part of an order may be superseded without affecting the rest of the order; but the annulling of the superseding order does not restore that part of the first order superseded.

A train order which has been annulled must not be reissued under its original number. If the former instructions are to be again made effective, it must be done under a new number.

Fifth. One order may qualify another without superseding it, for instance:

(a) A work extra instructed not to protect against extra trains may later be given an order to protect against certain extra trains.

(b) An order may direct a train to wait, or run late. Afterward another order may instruct them to wait longer, or run later. It is neither proper, nor necessary, to supersede a time order.

(c) A train may be ordered to display signals; and a later order may direct that signals be taken down.

(d) An order may direct an operator to "Hold Eastward Trains." A later order may instruct "No. 2 may go."

Sixth. If a crew holds an order to run extra A to C, they must pull into the siding at C, where all their orders become void, excepting slow orders and annulments. If later created as a new train, they must be provided with a new set of orders, excepting annulments and slow orders. Some roads permit the dispatcher, under such circumstances, to give the train an order, "Respect Order No. ....," when the crew has the order in their possession but cannot use it when recreated. This saves the dispatcher the time required to send them a new copy of the order.

When a work extra's working orders expire by limitation of time, all orders addressed to the work extra become void. The same applies to an extra whose running order is limited by time.

Seventh. If a train is given an order to run one hour late A to D, inferior trains may add the one hour to the train's time at A, B, and C; but not at D, nor at any point beyond D.

Eighth. Trains must be identified by the men in charge, in the absence of other positive evidence of identity. Identification is sometimes required to be made by deduction. If a train were waiting to meet No. 1 and a passenger train passed the station on No. 1's time without signals and having markers, it could be assumed to be No. 1, if no other westward schedule were overdue.

If the train were an extra, it would have white flags; and it could not move against the train except by train order.

Ninth. A regular train may be created by a clearance card at its initial station on each subdivision. It can only be created at an intermediate point by train order. When a schedule is given to a train by a clearance card, or otherwise, it retains the schedule to the end of its run on that subdivision; the exception being when the schedule is annulled by train order, or ceases to exist as provided by Rules 4 and 82.

Tenth. Any order not in proper form, or not understood, should be returned for correction, or annulment. Trains should never move on such orders.

FORM S-A

FIXING MEETING POINTS FOR OPPOSING TRAINS

Form A of the Standard Code does not show any form which authorizes the use of the term "take siding," or "hold main track." The rules of some roads are identical, which would seem to indicate that dispatchers are not permitted to issue other than straight meets, letting the inferior train take the siding.

Some roads authorize the use of the term "take siding" but do not permit the use of the term "hold main track." Others do it just the opposite; for instance, the Missouri Pacific does not permit the use of "take siding" in any order. The inferior train is always told to hold the main track, and this puts the superior train in the siding.

Here are some forms of S-A found in various rule books:

(a) No. 8 take siding meet No. 59 at B No. 65 at C and No. 151 at D.

No. 8 will take siding at each of the designated points for the trains named.

(b) No. 1 hold main track meet No. 2 at D No. 4 at D and No. 8 at B.

No. 1 will hold the main track at each of the designated points only for the trains designated.

(c) No. 15 get this order hold main track and meet No. 16 at C instead of D.

No. 16 must approach D carefully, taking into consideration ability of No. 15 to stop clear of switch to be used by No. 16 in going on siding.

(d) No. 15 hold main track meet No. 16 at C instead of D.

Order to No. 15 at C.

(Same instructions apply as under (c).

(e) No. 62 with .... cars back in and meet No. 61 at C.

This movement must be protected by a member of the crew of No. 62 as prescribed by Rule 99.

This last order, (e), will not be found in many rule books. It is claimed that the order is better than the old "pull by and back in" order; however, the majority of roads have dropped the form entirely. Sometimes it is necessary to have a train make a move, such as that for which the order provides; and if properly made, as prescribed by the rules, it is safer than letting the train pull by and back in on a straight meet. At least the opposing train knows what is going on.

A flagman should precede No. 62 and protect them while they are pulling up the main track to back in; and he should remain out to inform No. 61 that the move has been made. If No. 61 is not so informed, they should stop back far enough to allow a train with the number of cars stated in the order to pull by the siding switch. It is then difficult to ascertain at times whether the train is in the siding or has not yet arrived, which is the reason for the flagman to stay out and notify No. 61.

The more order forms a dispatcher has authorized, the more sensibly his orders should read. However, he should not mix them up in such a way as to confuse someone: for instance, an order like this.

ORDER NO. 2

> No. 1 take siding meet Extra 3433 East at D and hold main track and meet No. 2 at C and meet No. 242 at B.

Under such an order, No. 1 would take siding at D. hold main track at C; and, if No. 242 is of inferior class. No. 1 would hold main track at B; if No. 242 were of the same class, No. 1 would take siding at B. Such an order is confusing. If later the dispatcher superseded this order by instructing No. 1 to take siding and meet Extra 3433 East at C instead of D, No. 1 would then be in the predicament of having to take siding for Extra 3433 East and hold main track for No. 2 at the same station, C.

FORM B—DIRECTING A TRAIN TO PASS OR RUN AHEAD OF ANOTHER TRAIN

The Code does not authorize any "pass when overtaken" or "run ahead until overtaken" orders; yet, some roads do use these orders, as shown by the following examples:

(a) No. 6 pass No. 4 when overtaken.

Both trains will run according to rule until the second-named train is overtaken and then arrange for the rear train to pass promptly.

(b) Extra 95 West run ahead of No. 3 B until overtaken.

The first-named train will run ahead of the second-named train from the designated point until overtaken, and then arrange for the rear train to pass promptly.

The following order would seem to be an improvement over Standard Code Form B, example (3):

(3) No. 1 pass No. 3 at K and run ahead of No. 7 M to Z.

When an inferior train receives an order to pass a superior train, right is conferred to run ahead of the train passed from the designated point.

Outside of block system territory, the superior train must not exceed the authorized maximum speed of the inferior train ahead until the order is fulfilled. In such territory, this form of order will not be used, except under extraordinary circumstances.—Illinois Central.

An order to pass a train confers right on the passing train to continue ahead of the train passed to the end of its run, or the end of the subdivision.

At a passing point, the train to be passed should take the siding, when practicable. If the trains are given the order prior to their arrival at the passing point, it is well that the order state which train is to take the siding.

The rules of most roads prescribe that under a run-ahead order, the following train shall not exceed the authorized speed of the train ahead. The leading train should never assume that the order makes flagging unnecessary. No part of Rule 99 is suspended by Form B; neither are the provisions of Rule 93 nullified by Form B. A train must protect its rear against the train it is preceding.

If there are sections of a schedule and the dispatcher wishes a train to run ahead of each of them "until overtaken," the order must name each section.

If given an order to run ahead of No. 2, and the train is overtaken by First 2, the A. of A. R. has ruled that the train must then let all sections pass at that point. Here is the ruling:

Extra 1540 West received the following order: "Extra 1540 West run ahead of No 3 Engine 2189 B until overtaken." Extra 1540 West is overtaken by No. 3, displaying signals; has Extra 1540 West the right to proceed ahead of second No. 3?

When a train is named in a train order by its schedule number alone, all sections of that schedule are included. Therefore it is necessary for Extra 1540, after being overtaken by first No. 3, to remain on the siding until all sections of No. 3 have passed, unless otherwise directed.—November 21, 1912.

So many complications and misunderstandings and delays result from the use of Form B that the dispatcher should never use it when another order will serve the purpose, say a wait order, or run-late order. There are times when he positively must get one train to a given point ahead of another and cannot avoid the use of Form B;

such as when necessary to move a wrecker to a derailment ahead of a passenger train. Should trains bunch at a station, it is sometimes a "must" that an inferior train precede a more important train in order to break a blockade.

As a rule, Form B should only be used by the dispatcher as a last resort.

FORM S-C
GIVING RIGHT OVER AN OPPOSING TRAIN

It will be noted that the Standard Code does not provide any form which gives one extra right over another short of the end of its run. This is desirable; and the majority of roads provide this authority. There are also several forms authorized in various books which vary from the Code forms, such as:

(a) Eng. 37 run extra H to F with right over No. 3. The regular train must not go beyond the point last named until the extra train has arrived, unless authorized to do so by train order.—Western Pacific.

(b) Extra 72 West has right over Extra 91 East A to G but wait at E until nine fifty nine 9 59 a m and at F until ten fifteen 10 15 a m for Extra 91 East.

The first-named train must not pass the points designated as E and F before the time given at such points unless the other train has arrived, and must take siding and not pass the point designated as G unless the extra over which right has been conferred has arrived, or unless directed by train order to do so.—Santa Fe.

(c) Extra 35 West has right over No. 62 A to E and at E. Under this example Extra 35 West is superior to No. 62 at B, C, D and E. The order must be delivered to No. 62 before its arrival at E.—Western Pacific.

(d) Extra 600 South has right over Extra 601 North A to M and wait at F until 9 59 a m
G 10 39 a m
H 10 55 a m
for Extra 601 North and will not leave M unless Extra 601 North has arrived. Extra 601 North hold main track at M.—Uniform Code.

Various roads have rules similar to Example (b) of the Santa Fe. A short right-of-track between two extras, i.e., "Extra 227 West has right over Extra 228 East H to D," would be a perfectly good order on the Santa Fe. The rule requires that Extra 227 West take siding and not leave D unless Extra 228 East has arrived.

On roads which do not have such a rule in their books, this would constitute a lap of authority. The order would read: "Extra 227 West has right over Extra 228 East H to D take siding and not leave D unless Extra 228 East has arrived."

It will be noted that the Missouri Pacific, which does not permit the use of the term "take siding" have worded their order differently but the meaning is the same.

FORM D—GIVING RIGHT OVER ANOTHER TRAIN IN THE SAME DIRECTION

This form is sometimes a most useful tool in preventing delay to a following train of considerable importance, in order to have trains ahead clear their time, avoiding delay in passing.

The forms authorized for this order vary from one rule book to another. Here is an example:

(1) No. 9 has right over No. 19 A to Z.
The second-named train must clear the time of the first-named train as prescribed by Rule ....

(2) Extra 95 West has right over No. 59 A to Z.

(3) Extra 65 West has right over Extra 75 West A to Z.
Under examples (2) and (3) the second-named train must clear the train-order time, if any, of the first-named train as prescribed by Rule ....; otherwise, it must not leave A ahead of the first-named train.

FORM E—TIME ORDERS

Some roads do not permit the use of a run-late order in any form. Other roads permit, in addition to running a train late on its schedule, that a run-late order can be used in connection with a wait order. Here is an example:

ORDER NO. 17

No. 1 wait at N nine fifty nine 959 AM
P ten thirty 1030 AM
R ten fifty five 1055 AM

ORDER NO. 18

No. 1 run twenty 20 mins late on Order No. 17

Order No. 18 makes the time at N, P and R, twenty minutes later than the times shown in order No. 17, i.e., 10:19 a.m., 10:50 a.m. and 11:15 a.m.

It is usually permissible to vary this order so as to start the run late at an intermediate point, as: "No. 1 run twenty 20 mins late on Order No. 17 from P," which makes the times No. 1 must respect as follows: N 9:59 a.m., P 10:50 a.m. and R 11:15 a.m.

The run-late order is, as a rule, an undesirable order insofar as efficiency is concerned. It happens at times that a dispatcher is so busy he has no time to put out a long string of waits across a subdivision, based on the running time the train can make. He must choose between letting inferior trains take a delay because he has no time out on a first-class train, or running the first-class train late on its schedule, which does not allow the train to make up time.

In order to understand clearly how the "run-late" order works, let us assume the following schedule for No. 2:

SCHEDULE OF NO. 2

A	L	1001 AM
B		1015 AM
C		1025 AM
D		1040 AM
E		1050 AM
F		1105 AM
G		1115 AM
H	A	1130 AM

ORDER NO. 3

No. 2 run one hour late A to E

This order would make No. 2's time one hour later at A, B, C, and D; but does NOT affect No. 2's schedule time at E. If the dispatcher wishes, he can start a first section of No. 2 from E "on time." Therefore, a train at E could use 11:40 a.m. to go to D for No. 2; while an inferior train at F would have to clear the schedule time of 10:50 a.m. at E. It cannot use the 11:40 a.m. at D in order to move against No. 2 from F to E.

Now let us look at a time order, or wait order:

ORDER NO. 4

A	1101 AM
B	1115 AM
C	1125 AM
D	1140 AM

These waits restrict the movement of No. 2 exactly the same as Order No. 3; yet they have an entirely different effect on the situation. A dispatcher cannot start a first section of No. 2 from D, or any station between D and

H until after 11:40 a.m.; and an inferior train at F can use the 11:40 a.m. wait at D to move against No. 2 from F to E. If we run No. 2 late on order No. 4, it has a different effect from running No. 2 late on its schedule; for instance:

ORDER NO. 5

No. 2 run one hour late on Order No. 4

This order would make the times in Order No. 4 one hour later, including the time at D; and the dispatcher could not start a first section from D or any station beyond D until after 12:40 p.m.

These examples are worthy of careful consideration, since much confusion arises when men do not have a thorough understanding of the principles involved.

FORM G—EXTRA TRAINS

There is a wide variation in the forms authorized under Form G in the various rule books, many of which are not authorized in the Standard Code. The examples shown below were taken from the rules of different roads; however, no one road authorizes the use of all of them.

(a) Eng. 77 run extra leaving A Thursday Feb. 17th as follows with right over all trains.

Leave	A eleven ten	11 10 p m
	C twelve five	12 05 a m
	E one forty	1 40 a m
Arrive	F two twenty	2 20 a m

This order may be varied by specifying the kind of extra and the particular trains over which the

FORM 31 | EACH CONDUCTOR AND ENGINEER MUST HAVE A COPY OF THIS ORDER

Train Order No. 221 Dispatcher's Office July 8 1945

WESTERN PACIFIC

To C. & E. Eastward Trains Except First Class

At South Sacramento STATION	X	INITIALS Opr.; TIME	M

ENG 31 WORKS EXTRA SIX ONE 601 AM UNTIL SEVEN ONE 701 PM
BETWEEN OROVILLE YARD AND ELOCMER PROTECTING AGAINST
SECOND CLASS TRAINS NOT PROTECTING AGAINST EXTRA TRAINS
EXCEPT PROTECTS AGAINST
EXTRA 252 WEST AFTER TEN ONE 1001 AM
EXTRA 912 WEST AFTER ONE THIRTY 130 PM AND
EXTRA 323 WEST AFTER THREE ONE 301 PM

Repeated at 450 a M GSa INITIALS Chief Train Dispatcher

CONDUCTOR	TRAIN	MADE	TIME	RECEIVED BY
Jones	X 47	Com	812 a M	Jonesrand
			M	
			M	
			M	

extra shall, or shall not, have right. Trains over which the extra is given right must clear the time of the extra as prescribed by Rule .... Work extras, whether required to protect or not to protect against extras, must clear the time of this extra in the same manner.

(b) After two ten 2 10 p m Eng. 75 run extra A to H.
Engine 75 must not leave A as an extra before 2:10 p.m.

(c) After Extra 45 East arrives at G Eng. 50 run extra G to Z.
Extra 50 must not occupy the main track at G until after Extra 45 East has arrived there.

(d) Eng. 92 has until nine fifty 9 50 a m to run extra H to N.
This order is fulfilled and Extra 92 loses right at 9:50 a.m.

(e) Eng. 16 help Extra 52 West K to M then run extra M to K.
Extra 16 must not leave M until Extra 52 West has arrived there.

(f) Eng. 17 help No. 62 H to G clear Extra 312 West then run extra G to F with right over No. 61.
Extra 17 must not leave G unless Extra 312 West is at that point.

(g) Eng. 99 run extra B to C this order is annulled at 7 10 p m.
A train receiving this order must, at the time specified, keep clear of the main track or protect itself in both directions, as prescribed by Rule 99.

(h) On Oct. 17 after 6 45 a m Eng. 88 run extra Z to G.
This form to be used to authorize in advance a later

movement of an extra train from a point at which there is no means of communication or at which the office is closed.

The train must not leave the point first named before the time specified in the order.

The dispatcher finds that the more order forms he has available, the easier his work can be arranged. On the other hand, everyone must be familiar with the forms and know their exact meaning, or the good which the dispatcher realizes is offset through the hazard of misunderstandings. This is probably the reason why many roads hold their order forms to a minimum.

When the dispatcher starts an extra train, before authorizing its movement, he must protect the train against conflicting movements of other extras then on the railroad. One extra never knows when there is an opposing extra on the railroad, except when so informed by train order; and in the absence of some provision for the extras to meet, they will collide.

This protection must be provided before the running order for an extra is completed. If a dispatcher allowed opposing extras to move against each other without a meet, figuring on giving it to them later, any number of things might happen. He might drop dead and the man taking over would not catch the fact that the two extras had no meet. More likely, he would simply overlook issuing the meet order until too late.

To turn an extra train loose without having provided for its meeting every other extra then authorized on the road, constitutes a lap of authority.

In the handling of extra trains, attention is called again to the discussion on the definition of "extra trains" and "extras." It is absolutely essential that all concerned understand fully the terms authorized by their rules, and the precise meaning of the terms used on their road.

ORDER NO. 1

Eng. 41 meet four Extras 1 36 137 and 3 West at A and run extra A to H and has right over Extra 75 West A to F take siding not leave F unless Extra 75 West has arrived and wait at
D until 1159 AM
E until 1217 PM for Extra 75 West

ORDER NO. 2

Extra 37 East has right over Extra 38 West A to G and wait at
C until 959 AM
D until 1036 AM for Extra 38 West and will not leave G until Extra 38 West has arrived Extra 38 West hold main track at G.

ORDER NO. 3

Extra 77 West has right over Extra 78 East A to F.

ORDER NO. 4

Extra 37 East has right over Extra 38 West C to F.

On roads where there are a limited number of open offices, particularly at night, dispatchers are forced to give one extra train right over another, prescribing waits. The meet form is not always adequate. Often one extra is known to be moving, or ready to move, and is at a point where it can be given orders, while an opposing extra is "in the dark," that is, between open offices, where it is impossible to learn precisely when it will arrive at a given point. If overdue, there sometimes is no way of knowing why the delay, or its possible duration.

It will be noted that practices vary on different roads

as to the use of the words "and" and "but," "unless" and "until." Order No. 1 is seldom used, the majority of roads requiring that the order read: "After four Extras 1 36 137 and 3 West arrive at A . . ."; or, "After the arrival of four extras 1 36 137 and 3 West Eng 41 run extra A to H . . .," etc.

Under Order No. 1, the meet with any one of the four extras could be superseded, if later found necessary. Under the other forms given, Engine 41 does not have a running order until after the four extras have arrived at A; therefore, if it becomes desirable to move Extra 41 East from A for one of the extras, its running order must be validated by annulling the portion of the order reading: "After four Extras 1 36 137 and 3 West arrive at A."

The reason Order No. 1 instructs Extra 41 East to "take siding and not leave F unless Extra 75 West has arrived" is obvious. Extra 75 West holds authority to run extra from H to A. Since the rules do not require it to consider opposing extras unless instructed to do so by train order, it would approach F unaware that its authority to proceed had been restricted. It is therefore entitled to the use of the main track at F, where it receives the order.

Actually, this order is not well understood and the form has been abandoned by some roads because of hazards which have arisen.

Extra 41 East is actually instructed that, if they do not meet Extra 75 West before their arrival at F, they will take siding at F, whether or not Extra 75 West has arrived at F.

A case in point is where a dispatcher gave Extra 41 East a train order, prior to their arrival at F, stating that Extra 75 West had arrived at F. This was done because Extra 75 West was doing work on an obscured track where, in the dispatcher's opinion, Extra 41 East could not identify them.

Such an order did not authorize Extra 41 East to use the main track at F, as actually occurred. In the meantime, Extra 75 West finished their work on the obscured track and again occupied the main track between switches before Extra 41 East arrived and the two trains were involved in a "butting" collision which might have been extremely serious.

It must be understood that there is no superiority between extra trains except that conferred by train order. Under Order No. 1, if the dispatcher failed to instruct Extra 41 East to take siding and not leave F before Extra 75 West arrived, the result would be a lap of authority between F and H, as both extras hold authority to proceed between those points; and there would be nothing to prevent Extra 41 East from continuing on its run from F to H.

Order No. 2 shows how one road, under the same conditions, has reworded their order so as to eliminate the possibility of a misunderstanding as to which train is entitled to the main track at all points.

Under Orders 3 and 4, roads using these examples, coupled with time orders, have a rule in their book similar to the following:

> "This order gives right to the train first named over the other train between the points named.
> Neither train shall go beyond the point last named until the other train has arrived, unless directed by train order to do so. If the trains meet at either of the designated points, the first-named train must take the siding unless the order otherwise prescribes. If the trains meet between the designated points, the second-named train will take the siding, unless the order otherwise prescribes."

Inasmuch as this rule governs the movements of the

two extra trains where a short right-of-track order is issued, it is not necessary for the dispatcher to duplicate those instructions in the train order. It must be remembered, however, that unless the rule book contains instructions such as the above quoted rule, Orders 3 and 4 would constitute a lap of authority.

Order No. 4, if given to Extra 37 East at A, would constitute a lap of authority between A and C, since neither extra would be superior to the other. If Extra 38 West flagged from F to C, there would be nothing to prevent its continuing on its running order C to A. Extra 37 East must have arrived at C before such an order is proper; otherwise, it should read ". . . A to F."

FORM S-H—WORK EXTRA

Under the Code rules, a dispatcher cannot fix up a work train in a single order, if there are extras on the railroad which he wishes the work extra to protect against; that is, unless he wants them to protect against all from one, or both directions. He cannot say: ". . . not protecting against extra trains, except . . . etc."

He must issue a second order and instruct the work extra to protect against, or clear, those extras.

It is well to remember that, unless otherwise instructed, a work extra must protect against extra trains in both directions. Therefore, it is improper to tell a work extra to protect against an extra until after it has been told NOT to protect.

The majority of roads have worked out methods slightly different from the Standard Code for issuing their orders; and each man should examine his rule book carefully. To illustrate this point, shown below are the complete Form S-H instructions, as found in the Southern Pacific book of rules:

FORM S-H—WORK EXTRA

(1) Eng. 292 works extra six forty five 645 AM until five forty five 545 PM between D and E.

The work extra, whether standing or moving, must protect itself against extra trains within the work limits in both directions.

This may be modified by adding:

(2) Not protecting against Eastward extra trains.

The work extra must protect against westward extra trains.

---

(3) Not protecting against extra trains.

Protection against extra trains is not required.

---

(4) Not protecting against extra trains until two forty 240 PM.

Protection against extra trains is not required until 2:40 p.m., and extra trains must not enter the work limits before that time.

---

(5) Not protecting against extra trains except protects against Extra 73 East after ten thirty 1030 AM and Extra 85 West after one thirty 130 PM.

The work extra will not protect against the extra trains excepted until the times specified and the extra trains excepted must not enter the work limits before those times. Protection against other extra trains is not required.

Under examples (1), (2), (3), (4) and (5), the time of regular trains must be cleared.

When a work extra has been instructed by train order not to protect against extra trains, and later it is desired that after a certain time it clear the track for, or protect against, a designated extra, an order may be given as in the appropriate example of the following form:

(6) Work Extra 292 clears extra 76 East between D and E after two ten 210 PM.

(7) Work Extra 292 protects against Extra 76 East between D and E after two ten 210 PM.

Under examples (6) and (7) Extra 76 East must not enter the work limits before 2:10 p.m.

To enable a work extra to work on the time of a regular train, the following form will be used:

(8) Work Extra 292 protects against No. 55 (or .... class trains) between D and E.

The work extra may work on the time of the train or trains mentioned in the order, and must protect against such train or trains.

When a work extra is to be given exclusive right over all trains, the following form will be used:

(9) Work extra 292 has right over all trains between D and E seven fifteen 715 PM until one fifteen 115 AM.

This gives the work extra the exclusive right to the track between the points designated between the times named.

This order must not be modified or amplified, and must not be issued while an order, example (3), Form G is in effect. (Schedule order.)

The work limits should be as short as practicable, to be changed as the progress of the work may require.

Work extras must give way to all trains as promptly as practicable.

> Whenever extra trains are run over work limits they must be given a copy of the work order sent to the work extra. Should the work order instruct a work extra not to protect against extra trains in one or both directions, extra trains must protect against the work extra; if the order indicates that the work extra is protecting itself against other trains, such trains must run expecting to find the work extra protesting itself.—Southern Pacific Lines.

If an extra train is to be stopped and used for work service for a short time, the extra's running order should be annulled before it is given a work order.

If a regular train is thus stopped, it is not necessary that their schedule be annulled; however, they should leave their train at the point where they enter the work limits, returning for the train when through working. They must not, for instance, take their train while working between C and D, then resume their schedule at D when the work is completed.

If a work extra is sent outside its limits for any considerable period of time, its working order should be annulled and a new one issued when it returns.

In addition to showing work trains in the space provided for on the train sheet, the dispatcher should indicate their working limits in the station column in some manner to prevent his overlooking the movement. Some roads require that the work extra be entered on either side of the train sheet in the train column, although dispatchers are not required to make a record of the work extra's movement in the OS spaces.

A meet order is not a good order to use in moving a work extra, as is reflected by the following A. of A. R. ruling:

A train order has been sent, reading:

"Engine 292 works 7:00 a.m. to 2:00 p.m. between

D and E. Work extra 292 south will meet No. 328 at E."

At 2:00 p. m. work extra 292 has not arrived at E. Does the meet order held by 328 hold good after 2:00 p. m., or can 328 proceed on its schedule?

If an order should be sent in the form suggested in the inquiry, it would be the duty of No. 328 to remain at E until Work Extra 292 arrived or until the order was annulled. Form A, fixing meeting point for opposing trains, states definitely that trains receiving these orders will run with respect to each other to the designated points, and there meet in the manner provided by the rules. If it is the intention of the dispatcher that No. 328 should move after 2:00 p.m. if the work extra had not arrived at that time, another form of order should be used.—October 17, 1910.

Under the circumstances the third form of Form E of Train Orders would have been the preferable form to use.—September 7, 1893.

FORM D-H—WORK EXTRA

Once the work train order has been mastered on single track, there is no mystery about the work extra on two or more tracks. However, given below is one A. of A. R. ruling in which the student will be interested:

The question has been asked whether a work extra working on one or both tracks on double track without being required to protect against extras has the right to move against the current of traffic or only to move with the current of traffic. (D-Form H, second example.)

This order gives the extra ample authority to move against the current of traffic within the limits and time named, properly clearing the time of regular trains.—October 17, 1910.

FORM J—HOLDING ORDER

The holding order is addressed only to the operator. As each train arrives and is held, it must be delivered a copy of the order, and at such time as each train is released, it must be given a copy of the order that releases it. This may be done in two ways: the trains may be released individually by an order "No. 2 may go"; or the dispatcher may annul the holding order to the operator. In either event, the two orders must be delivered to each train, together with a clearance card. Some of the conditions requiring the use of Form J are: Derailments, washouts, snowslides, burned-out bridge, or other interruptions of a nature which make it uncertain as to when traffic may be resumed. This form is not intended for use in moving one train against another.

FORM K

ANNULLING A SCHEDULE OR SECTION

About the only deviation from the form given in the Standard Code is an order which some rule books provide for, reading as follows:

> "Second 15 due to leave E Feb. 19th has arrived at E with no signals and is annulled E to Z."

It is not good practice to combine Form K with orders containing other movements which the dispatcher may later wish to annul. A schedule once annulled cannot be restored by annulling the order which annuls the schedule. However, if such an order is annulled, there is a question in the minds of the crew as to whether they still have the annulment of the schedule. The rules of most roads provide that signals must not be taken down at an intermediate point of a schedule other than a register station and that, where a last section is discontinued, it must be done by the use of Form K, allowing the preceding section to carry signals into the terminal.

FORM L—ANNULLING AN ORDER

The dispatcher may annul any order which he does not wish delivered, to the operator, who will write on his copy "annulled by Order No. ...." and file it. The same form is used when annulling the authority of an extra. In this case, the order is addressed to the train which holds the order to be annulled.

Dispatchers should avoid delivering a train order to a train showing that an order is annulled, if the train does not have the order annulled. To do so would create a doubt in the minds of the crew as to the contents of the order which had been annulled. It is only reasonable for the crew to demand a copy of the order annulled, since it may have been an order which had not been delivered to them which superseded some instructions in the orders they hold.

There was a case where a dispatcher issued Order No. 47 which ran a passenger train forty minutes late. Later, the dispatcher was informed that the passenger train would be ready to leave on time and he issued Order No. 52 to a freight crew annulling Order No. 47. The conductor misread the order number as Order No. 49 which was annulled and used the forty minutes time on the passenger train, resulting in a collision.

Not having Order No. 49, the crew dismissed the matter from mind on the supposition that the order, whatever it may have been, was annulled anyway and that there was no use asking about it. Had they requested a copy of the order annulled, this collision would have been averted.

FORM S-M
ANNULLING PART OF AN ORDER

There is a difference of opinion as to the proper use of Form S-M, some roads contending that only those words which are to be lifted out of the order, so to speak, should be quoted in a Form S-M order. This is contrary to a long-established practice and has been ruled improper by the Rules Committee of the A. of A. R. as shown below:

> Question: With reference to Train Order Form S-M of the Standard Code, will you please give me the following information?
>
> **ORDER NO. 10**
>
> No. 1 meet No. 4 at D and No. 2 at C.
>
> **ORDER NO. 11**
>
> That part of Order No. 10 reading, No. 1 meet No. 2 at C, is annulled.
>
> **ORDER NO. 12**
>
> That part of Order No. 10 reading, and No. 2 at C, is annulled.
>
> 1. What is your decision as to proper wording; should it be as shown in Order No. 11, or as shown in Order No. 12?
> 2. If the wording in Order No. 12 is improper, just what is the reason?
> 3. If the wording in Order No. 11 is proper, then does it, in fact, leave the remaining words of Order No. 10 reading: "No. 4 at D and"?
>
> Answer: Order No. 11 is shown as prescribed in the Standard Code. Order No. 12 is not covered by the Code. The wording Order No. 11 is proper and annuls the meet of No. 2 at C and allows the meet of No. 4 to stand.—Sept. 27, 1937.

FORM P
SUPERSEDING AN ORDER OR PART OF AN ORDER

An order may be superseded by adding to prescribed forms the words "instead of." One order must never be understood to supersede another with which it conflicts unless the words "instead of" are added. An order which has been superseded must not be re-issued under its original number. A part of an order may be superseded without in any way countermanding the remaining instructions.

It may be used in many ways by a dispatcher and is one of his most useful orders. For instance, should he be furnished a wrong engine number which has been included in an order, he may issue an order like this:

"Order No. 75 should read No. 2 Engine 484 instead of Eng 485."

Otherwise he might have to take up the erroneous order and reissue a new one, which might have complications if it had been delivered to other trains. Corrections in almost any part of an order may be made in the same manner.

However, there are certain limitations to the use of Form P. Forms of orders should not be mixed in a superseding order, for instance:

"No. 1 has right over No. 2 D to H instead of meet No. 2 at F."

The Association of American Railroads has ruled thus:

"Train orders issued under Forms A, B, C and F of the Standard Code may be superseded, but if it is desired to change orders issued under other forms, they should be annulled.—April 4, 1917.

Therefore, it is not proper to supersede a time order. It is not necessary if the time is to be increased. If necessary to reduce the time in a run late, or wait order, the order must be annulled and a new one issued.

MISCELLANEOUS

The following order forms do not appear in the Standard Code, but are found in similar form in many rule books and are reproduced for the information of the student, who may find similar forms in his book of rules.

FORM D-S

PROVIDING FOR THE USE OF A SECTION OF DOUBLE TRACK AS SINGLE TRACK

(1) No. 1 (or westward) track will be used as single track between F and G.

If it is desired to limit the time for such use, add:

FROM ONE ONE 1 01 PM TO
THREE ONE 3 01 PM

All trains must use the track specified between the stations named and must be governed by rules for single track.

Train running against the current of traffic on the track named must be clear of the track at the expiration of the time specified, or protected as prescribed by Rule 99.

Where trains moving with the current of traffic are authorized to run extra or work extra without train-order authority, Form D-S must not be used until all trains and detached engines, whether in service or tied up, which may use either track, have received a copy of the order.—Southern Pacific.

FORM V

CHECK OF TRAINS

(1) .... TRAINS DUE .... BEFORE ......
HAVE ARRIVED AND LEFT EXCEPT .....

(2) NO. .... OF .... HAS ARRIVED (or left)

(3) EXTRA .... HAS ARRIVED (or left) ......
ON ORDER NO. ....

When necessary to make an exception of a train or trains, such exception will be added.

Where not required, the word "arrived" or "left" will be omitted.

Form V must not be combined with other forms of train orders. When this form is received trains may register by ticket, provided it is not necessary to check the register against other schedules.—Southern Pacific.

Note: This form varies with different carriers. Some require that the date be included. A check may be given on "Regular Trains," "First Class Trains," "Superior Trains," etc. When a train-order check is given on "overdue superior trains," those trains which may have been given right over the train must be considered. It is interesting to note that some roads do not authorize the use of this type of order.

FORM W—CHANGE IN CLEARANCE OR REGISTER REQUIREMENTS

(1) .... NEED NOT OBTAIN CLEARANCE AT ....

(2) .... NEED NOT REGISTER AT ....

(3) Extra .... REGISTER AT .... ON ORDER NO. ....

(4) .... MAY CHECK REGISTER AT .... AGAINST EXTRA ..... ON ORDER NO. ....
—Southern Pacific.

Part III

QUESTIONS AND ANSWERS

What is of the first importance in the discharge of duty?

Safety.

What is essential to safety?

Obedience to the rules.

What assurance is given when entering or remaining in the service?

Willingness to obey the rules.

What does the service demand?

The faithful, intelligent and courteous discharge of duty.

What must be shown to obtain promotion?

Ability for greater responsibility.

Who must provide themselves with a copy of the rules?

Employes whose duties are prescribed by them.

Who must have a copy of the current time-table with them while on duty?

Employes whose duties are in any way affected by it.

What is required of employes with respect to the rules and special instructions?

They must be conversant with and obey them.

If in doubt as to their meaning what must they do?

Apply to proper authority for an explanation.

Who must pass the required examinations?

Designated employes.

Are persons employed in any service on trains subject to the rules and special instructions?

Yes.

What is required of employes with respect to carrying out the rules and special instructions?

They must render every assistance in their power, and report to the proper official any violation thereof.

When, how, and to whom must accidents, failure in the supply of water or fuel, defects in the track, bridges, signals or any unusual conditions which may affect the movement of trains be reported?

Promptly by wire to the proper authority.

Is the use of intoxicants or narcotics prohibited?

Yes.

When is the use of tobacco, by employes, prohibited?

While on duty in or about passenger stations, or on passenger cars.

What are the requirements as to badges, uniforms and appearance?

Employes on duty must wear the prescribed badge and uniform and be neat in appearance.

To avoid annoyance to the public, what are the requirements of employes, and others authorized to transact business at stations and on or about trains?

They must be courteous, orderly and quiet.

What is required of employes in case of danger to the Company's property?

They must unite to protect it.

What is required of employes to avoid injury to themselves or others?

They must exercise care and observe the condition of equipment and the tools which they use in performing their duties, and when found defective will, if practicable, put them in safe condition, reporting defects to the proper authority.

Must employes know the location of structures or obstructions where clearances are close?

Yes.

When are trains expected to be run?

At any time, on any track in either direction.

Is it prohibited to stand on the track in front of an approaching engine or car for the purpose of boarding the same?

Yes.

What is an Engine?

A unit propelled by any form of energy and used in train or yard service.

What is a Train?

An engine or more than one engine coupled, with or without cars, displaying markers.

What is a Regular Train?

A train authorized by a time-table schedule.

What is a Section?

One of two or more trains running on the same schedule displaying signals or for which signals are displayed.

What is an Extra Train?

A train not authorized by a time-table schedule.

How may any extra train, except passenger extra or work extra, be designated?

As extra.

How may passenger train extra be designated?

As passenger extra.

How may work train extra be designated?

As work extra.

What is a Superior Train?

A train having precedence over another train.

What is a Train of Superior Right?

A train given precedence by train order.

What is a Train of Superior Class?

A train given precedence by time-table.

What is a Train of Superior Direction?

A train given precedence in the direction specified by timetable as between opposing trains of the same class.

What is a Time-Table?

The authority for the movement of regular trains subject to the rules. It contains classified schedules with special instructions relating to the movement of trains.

What is a Schedule?

That part of a time-table which prescribes class, direction, number and movement for a regular train.

What is a Division?

That portion of a railroad assigned to the supervision of a superintendent.

What is a Subdivision?

A portion of a division designated by time-table.

What is a Main Track?

A track extending through yards and between stations, upon which trains are operated by time-table or train order, or both, or the use of which is governed by block signals.

What is a Single Track?

A main track upon which trains are operated in both directions.

What are Two or More Tracks?

Two or more main tracks upon any of which the current of traffic may be in either specified direction.

What is Current of Traffic?

The movement of trains on a main track, in one direction, specified by the rules.

What is a Station?

A place designated on the time-table by name.

What is a Siding?

A track auxiliary to the main track for meeting or passing trains.

What is a Fixed Signal?

A signal of fixed location indicating a condition affecting the movement of a train or engine.

(See the Standard Code note to Definition of Fixed Signal.)

What is a Yard?

A system of tracks within defined limits provided for the making up of trains, storing of cars and other purposes, over which movements not authorized by time-table, or by train order, may be made, subject to prescribed signals and rules, or special instructions.

What is a Yard Engine?

An engine assigned to yard service and working within yard limits.

What is a Pilot?

An employe assigned to a train when the engineman or conductor, or both, are not fully acquainted with the physical characteristics or rules of the railroad, or portion of the railroad, over which the train is to be moved.

What is a Train Register?

A book or form which may be used at designated stations for registering signals displayed, the time of arrival and departure of trains and such other information as may be prescribed.

What is Reduced Speed?

Proceed prepared to stop short of train or obstruction.

What is Yard Speed?

A speed that will permit stopping within one-half the range of vision.

How are rules for single track identified?

By the prefix "*S.*"

How are rules for two or more tracks identified?

By the prefix "*D.*"

To what do rules without a prefix apply?

To single and two or more tracks.

If an engine is met with markers displayed but without cars, has a train been met?

Yes.

If an engine and cars are met, engine at the front, caboose at rear, but no markers displayed, has a train been met?

No.

If cars are met with markers displayed on the rear car but without an engine attached, has a train been met?

No.

Except when restricted in some manner for the inferior train, is the superior train required to know that the inferior train is complete?

No.

Have extra trains any authority except that which is conferred by train order?

No, not unless otherwise provided.

May extra trains be described in train orders as passenger, fruit, motor, etc.?

"Passenger" is the only designation prescribed by the Standard Code; however, it is the practice on many roads to use others.

As used in these rules, what is meant by "precedence"?

Superiority-authority to run ahead of, or against, another train or trains.

If a train were ordered to run extra on double track, which track must it use?

It must run with the current of traffic.

How must a work extra move on two or more tracks?

It must move with the current of traffic.

Do yard engines require train orders to work within yard limits?

No.

Where must the pilot of a train ride?

On the engine.

On which pages of a train register should regular and extra trains register?

Some roads require that regular trains (including all sections) be registered on the page corresponding with the date the train is due at the stations where registers are provided; and that extra trains register on the page corresponding with the date it actually arrives or leaves. Other roads require that all trains register on the page corresponding with the date they actually arrive or leave.

What is Remote Control?

A term applied to a system of operating outlying switches or signal appliances from a designated point.

What is a Dual Control Switch?

A power-operated switch which is also equipped for hand operation.

What is a Spring Switch?

A switch equipped with a spring, so that when run through in trailing movements, the switch points return to their original position.

What is a Cab Signal?

A signal located in engineer's compartment or cab, indicating a condition affecting the movement of a train or engine and used in conjunction with interlocking signals and in conjunction with or in lieu of block signals.

What is an Absolute Signal?

A block or interlocking signal designated by an "A" marker, or by the absence of a number plate.

What is an Absolute Block?

A block in which but one train or engine at a time is permitted.

What is a Control Station?

A place from which remote control switches or signal appliances are operated.

What is a Control Operator?

An employe in charge of a control station.

When so prescribed, will the train dispatcher perform the duties of control operator?

Yes.

When will Standard Time be transmitted to all points from designated offices?

12:00 noon Eastern Time, 11:00 a.m. Central Time, or 10:00 a.m. Mountain Time, daily.

What will govern time service?

Instructions issued by proper officer.

By whom must watches that have been examined and certified to by a designated inspector be used?

By such officers and employes as are designated by instructions governing time service.

When must officers and employes required to use standard watches submit their watches for inspection, unless excepted by special instructions?

As often as required by the individual rule book.

Must they sign inspection record?

Yes.

How will it be known where standard clocks are located?

They will bear the sign "Standard Clock" and their locations will be shown by special instructions.

Where standard clocks are available, when must the watches of employes be compared with a standard clock.

Before commencing each day's work or trip.

Must the time when watches are compared be registered on a prescribed form?

Yes.

How must employes whose duties preclude access to a standard clock obtain time or compare and regulate their watches?

They must obtain time from train dispatcher, or compare and regulate their watches daily with those of conductors and engineers who have standard time and have registered as provided.

When must conductors, or engine foremen, and engineers

also compare time with each other, and with trainmen, or yard-men, and firemen?

Before commencing each day's work or trip, when practicable.

For the purpose of operation and to avoid possible confusion, must each subdivision (or district) be regarded as a separate unit of railroad?

Yes.

How many schedules of the same number and day shall be in effect on any division or subdivision?

Not more than one.

Should footnotes be found on certain pages of time-tables, to what do they apply?

Only to the page on which they appear.

Should time-table special instructions be found to conflict with the Book of Rules, which will govern?

Time-table special instructions.

Should bulletin instructions or special notices be issued superseding or modifying any of the foregoing, what must be understood?

Bulletin instructions or special notices take precedence.

Should a train order conflict with special instructions issued in the form of a bulletin or by time-table, which governs?

The train order.

TIME-TABLES

When does a new time-table supersede the preceding time-table?

From the moment it takes effect.

When Rule 4 refers to a schedule of a preceding (old) time-table, does it mean one that is due or overdue, not exceeding twelve hours at the time of change?

Yes, one that actually exists and authorizes a train.

When does a schedule exist?

When it becomes due.

When does a schedule of the new time-table first exist if no schedule corresponds with it on the old time-table?

At its first leaving time at its initial station, on any division or subdivision, at or after the new time-table takes effect.

In what manner must a schedule of the old time-table correspond with a schedule of the new time-table to permit a train authorized by the old time-table to assume a schedule of the new time-table?

It must correspond in class, number, direction, route, day of leaving, initial and terminal stations.

Note: Some lines do not include "route" as one of the requirements, interpreting "direction" to include "route."

What is meant by "day of leaving"?

"Day of Leaving" refers to the heading of the schedule column in the time-table where, under the number of the schedule, usually appears the wording "Daily" or "Daily except Sunday," etc. If "Daily" on the old time-table and "Daily" on the new time-table, they would correspond as to the day of leaving any day of the week. If "Daily' on the old time-table and "Daily except Sunday" on the new time-table, or vice versa, they would not correspond as to the day of leaving on Sunday but would correspond every other day of the week.

If "Daily except Sunday" on the new time-table and "Daily except Saturday" on the old time-table, could they be considered as corresponding as to the day of leaving on Saturday and Sunday; and how would it affect the schedule on Monday, Tuesday, Wednesday, Thursday or Friday?

They would not correspond on Saturday and Sunday; but would correspond other days of the week.

Is it necessary that the schedule of the old and new time-table correspond in "day of arriving."

No.

If a schedule of the old time-table corresponds in the seven requirements with a schedule of the new time-table, how will a train authorized by, or a train fulfilling a schedule of the old time-table proceed when the new time-table takes effect?

It will retain its train orders and assume the schedule of the corresponding number in the new time-table.

If a schedule of the old time-table does not correspond in

either or all of the seven requirements with a schedule of the new time-table, how will a train authorized by, or a train fulfilling a schedule of the old time-table proceed when the new time-table takes effect?

Only as authorized by train order.

If a schedule of the old and new time-tables are alike in Number, Class, Day of Leaving, Initial and Terminal Stations, but do not follow the same route, would they correspond? Why?

No. They would differ in route.

Note: On roads where route is not included as one of the requirements, the answer would be: No. They would differ in direction. Direction means they must follow the same route.

How should a train running under the old time-table be governed after the new time-table takes effect if the route (direction) is not the same?

Take siding and proceed only as authorized by train order.

Where does a schedule take its date?

From its initial station on any division or subdivision.

Note: Some roads use the term "district" instead of "subdivision."

If No. 2 due to leave A at 11:30 p.m. January 1st, does not leave until 1:00 a.m. January 2nd, what is its date?

January 1st.

What is the difference between a time-table schedule and a train that runs under its authority?

A schedule is that part of a time-table which authorizes the existence of a regular train; and the train is the equipment that moves on such authority.

May a train be authorized under the old time-table (meaning a schedule in effect) without the equipment being on the road?

Yes.

Is it necessary to have an authorized train or a schedule due, or overdue, not to exceed twelve hours, at some station on the old time-table at the time of change before a schedule of the new time-table could be assumed?

Yes.

If there is a train authorized by the old time-table at the time of change, is it necessary to also have a train authorized by the new time-table at the time of change in order that the schedule of the corresponding number may be assumed?

No.

If the schedule of the new time-table, corresponding as required, is not due or overdue at the time of change, when does it first exist?

When it becomes due.

May more than one schedule of the same number and day be in effect on any division, or subdivision?

No.

NEW TIME-TABLE EFFECTIVE 12 01 A M SUNDAY

OLD				NEW		
WESTWARD				WESTWARD		
	1st CLASS				1st CLASS	
	1				1	
	DAILY				DAILY	
A	L	10 00 PM		A	L	11 00 PM
B		10 40 PM		B		11 40 PM
C		11 20 PM		C		12 20 AM
D		12 01 AM		D		1 00 AM
E		12 50 AM		E		1 50 AM
F		1 40 AM		F		2 40 AM
G		2 20 AM		G		3 20 AM
H	A	3 00 AM		H	A	4 00 AM

Do the schedules correspond in the seven requirements?

Yes.

Is there a train authorized (meaning a schedule in effect not fulfilled) on the old time-table at time of change, 12:01 a.m. Sunday? Where?

Yes, at D.

Is there a schedule due or overdue on the new time-table at the time of change?

Yes.

At 12:01 a.m. Sunday, when must No. 1 of the new time-table be considered as having taken effect?

At 11:00 p.m. Saturday, one hour and one minute before the change.

Must No. 1's time on the old time-table be exactly 12:01 a.m. at some particular station in order that a train may be authorized?

No. If due at D at 11:50 p.m. and not fulfilled at D at 12:01 a.m., a schedule would exist as it is not 12 hours late at the time of change.

If No. 1 has not left B prior to 12:01 a.m. Sunday, would there be a train authorized at B? Why?

Yes, because the schedules correspond as required and there is a schedule due and not fulfilled.

If a No. 1 of Saturday fulfills the schedule of the old time-table from A to D, reaching D at 12:01 a.m. Sunday, what effect would this have on the corresponding portion of the schedule of the new time-table?

The corresponding portion of the new schedule of the No. 1 of Saturday between A and D could not be used.

If No. 1's schedule of the old time-table is not fulfilled out of A before 12:01 a.m. Sunday, over what part of the Division, or Subdivision, is there a train authorized at the time of change?

From A to D.

As the new time-table takes effect at 12:01 a.m. Sunday, what is the date of the last No. 1 existing out of A under the old time-table?

No. 1 of Saturday.

How many No. 1's (ones) of Saturday may exist?

Only one.

Is it necessary only to have a train authorized by the old time-table somewhere on the Division or Subdivision at 12:01 a.m. Sunday, in order that the schedule of No. 1 of the new time-table may be assumed?

Yes.

If a train inferior to and moving in the same direction

as No. 1 could clear No. 1's schedule time at D on the new time-table, could it leave ahead of, or pass No. 1 without orders?

Yes. It could pass No. 1 at D between 12:01 a.m. of the old time-table and 1:00 a.m. of the new time-table.

If an opposing inferior train leaves H Saturday and reaches E at 11:45 p.m., how much time would it then have to go to D on the old time-table?

Sixteen minutes, less required clearance.

If it is a 30-minute run E to D when could this opposing inferior train leave E on new time-table, and how much time would it then have to go to D?

At 12:01 a.m. Sunday, and would have fifty-nine minutes, less required clearance to go to D against the new time-table.

Would it be safe to assume that as No. 1 could not leave D before 12:01 a.m. Sunday (the time of change), that the inferior train could leave E at 11:45 p.m. and move against the new time at D (1:00 a.m.)?

No.

If No. 1 leaves A Saturday on the old time-table and reaches C at 11:50 p.m. (11 minutes before the change), assuming that it is a 20-minute run C to D, how should it be governed?

Remain at C under the old time-table until 12:01 a.m. Sunday; then No. 1 is not due to leave C until 12:20 a.m., at which time it could proceed.

In what position would No. 1 of Saturday be if it left C at 11:10 p.m. and was unable to make D by 12:01 a.m. Sunday?

It would be between C and D when the new time-table took effect, 19 minutes before due out of C, the next station in the rear, on the new time-table.

If opposing inferior train at D is unable to make C for No. 1 on the old time-table, could it move against No. 1 on the new time-table leaving D at 12:01 a.m. Sunday, provided it can clear the new time of No. 1 at C (12:20 a.m.) as required by rule?

Yes.

What takes the place of that portion of the old time-table D to H?

That portion of the schedule of the new time-table.

What would No. 1, arriving at D at 12:01 a.m. Sunday, do?

No. 1 would remain at D until 1:00 a.m., Sunday, then proceed.

If No. 1 received an order to meet No. 2 at G, how should it be governed if there is no No. 2 on the new time-table?

The same as if the order had not been issued.

If No. 1, while running on the old time-table, received an order to meet an opposing extra at G, would it respect the meet with the extra at G if it assumes its time on the new time-table at or before arrival at G, the same as it would have done had there not been a change of time-table?

Yes.

When would No. 1 of Saturday become twelve hours late at H?

4:00 p.m. Sunday.

NEW TIME-TABLE EFFECTIVE 12 01 A M SUNDAY

OLD		
WESTWARD		
	1st CLASS	
	1	
	DAILY	
A	L	10 00 PM
B		10 40 PM
C		11 20 PM
D		12 01 AM
E		12 50 AM
F		1 40 AM
G		2 20 AM
H	A	3 00 AM

NEW		
WESTWARD		
	1st CLASS	
	1	
	DAILY	
A	L	9 00 PM
B		9 40 PM
C		10 20 PM
D		11 00 PM
E		11 50 PM
F		12 40 AM
G		1 20 AM
H	A	2 00 AM

If No. 1, the time of which on the new time-table is earlier than on the old time-table, leaving A Saturday on the old time-table, arrives at C at 11:50 p.m., could it proceed at 11:50 p.m.? Why?

Yes, because the schedule on the new time-table is earlier

than the schedule of the old time-table; and No. 1 would proceed as a delayed train.

Would it, then, be proper for No. 1 of Saturday to be between C and D at 12:01 a.m. Sunday, provided it was not on the time of an opposing superior schedule of the new time-table?

Yes.

Would No. 1 be permitted to make up time on the new time-table if no order to prevent its doing so?

Yes.

Could No. 1 of Saturday pass D on time at 12:01 a.m. Sunday?

No, the minute No. 1 is due to leave D on the old time-table, the new time-table takes effect, which would make it one hour and one minute late.

If an opposing inferior train is at F at 11:40 p.m. Saturday, on the old time-table, how much time would it have to make E, and why?

Twenty-one minutes. At 11:40 p.m. at F, it has until 12:50 a.m. to make E as long as the old time-table is in effect; but as the new time-table takes effect at 12:01 a.m., No. 1's time at E is then 11:50 p.m., instead of 12:50 a.m.; and it only has until 12:01 a.m. to make E.

NEW TIME-TABLE EFFECTIVE 12 01 A M SUNDAY

OLD		
WESTWARD		
	1st CLASS	
	3	
	DAILY	
A	L	10 00 PM
B		10 40 PM
C		11 20 PM
D		12 01 AM
E		12 50 AM
F		1 40 AM
G		2 20 AM
H	A	3 00 AM

NEW		
WESTWARD		
	1st CLASS	
	13	
	DAILY	
A	L	9 00 PM
B		9 40 PM
C		10 20 PM
D		11 00 PM
E		11 50 PM
F		12 40 AM
G		1 20 AM
H	A	2 00 AM

Do the schedules correspond in the seven requirements?

No.

In what do they fail to agree?

Number.

As there is not a schedule of the new time-table with which No. 3 of Saturday on the old time-table corresponds in the seven requirements at the time of change, what becomes of the schedule of No. 3 of the old time-table?

It ceases to exist.

How may a train running as No. 3 of Saturday of the old time-table proceed after the new time-table takes effect?

Only as authorized by train order.

As there is not a No. 13 of Saturday on the old time-table corresponding in the seven requirements with No. 13 of the new time-table, when is No. 13 of the new time-table first authorized to run?

At 9:00 p.m. Sunday, which is the first leaving time at its initial station after the new time-table takes effect at 12:01 a.m. Sunday.

NEW TIME-TABLE EFFECTIVE 12 01 A M SUNDAY

OLD		NEW		
WESTWARD		WESTWARD		
			1st CLASS	
			5	
			DAILY	
		A	L	10 00 PM
		B		11 00 PM
		C		12 15 AM
		D		1 30 AM
		E		2 35 AM
		F		3 45 AM
		G		4 50 AM
		H	A	6 00 AM

As there is no No. 5 on the old time-table, when does No.

5 of the new time-table first exist, or when is it first authorized to run?

10:00 p.m. Sunday, at its initial station, which is the first leaving time following the time of change.

As there is no schedule on the old time-table, could No. 5 of the new time-table become authorized at any intermediate station at the first leaving time after the new time-table takes effect? For example, at C 12:15 a.m. Sunday?

No.

If the new time-table should take effect at 10:00 p.m., Sunday, instead of 12:01 a.m. Sunday, when is No. 5 of the new time-table first authorized to run?

10:00 p.m. Sunday.

If the new time-table should take effect at 10:01 p.m. Sunday, instead of 12:01 a.m. Sunday, when is No. 5 of the new time-table first authorized to run?

10:00 p.m. Monday.

NEW TIME-TABLE EFFECTIVE 1 30 A M SUNDAY

OLD			NEW		
WESTWARD			WESTWARD		
	1st CLASS			1st CLASS	
	7			7	
	DAILY			DAILY	
A	L	10 00 PM	A	L	2 00 AM
B		11 00 PM	B		3 00 AM
C		12 15 AM	C		4 15 AM
D		1 30 AM	D		5 30 AM
E		2 35 AM	E		6 35 AM
F		3 45 AM	F		7 45 AM
G		4 50 AM	G		8 50 AM
H	A	6 00 AM	H	A	10 00 AM

As new time-table takes effect at 1:30 a.m. Sunday, what is the date of the last No. 7 leaving A on the old time-table?

No. 7 of Saturday.

What is the date of the first No. 7 that can exist on the new time-table?

No. 7 of Sunday.

What becomes of No. 7 of Saturday of the old time-table at the time of change, (1:30 a.m. Sunday)?

It ceases to exist.

If an extra, or other inferior train, moving in the opposite direction to No. 7, Saturday night, receives an order to meet No. 7 at B, and it is at F at 1:30 a.m. Sunday, when the new time-table takes effect, could the meet be used to go to B for No. 7 on the new time-table? Why?

No. because No. 7 of the new time-table is No. 7 of Sunday and the order was to meet No. 7 of Saturday of the old time-table.

When an order is received to meet a regular train, to what regular train does it refer?

To the regular train of that number specified in the order then due on the division, or subdivision; and if none due, then to the first of that number that will be due.

After 6:00 p.m. Saturday, the time that No. 7 of Friday becomes twelve hours late at H, to what train would an order that is received any time before 1:30 a.m. Sunday refer?

No. 7 of Saturday.

After 1:30 a.m. Sunday, how must inferior trains be governed regarding No. 7 of the new time-table?

Clear No. 7 of the new time-table as required by rule.

NEW TIME-TABLE EFFECTIVE 2 30 A M SUNDAY

OLD		
WESTWARD		
		1st CLASS
		7
		DAILY
A	L	10 00 PM
B		11 00 PM
C		12 15 AM
D		1 30 AM
E		2 35 AM
F		3 45 AM
G		4 50 AM
H	A	6 00 AM

NEW		
WESTWARD		
		1st CLASS
		7
		DAILY
A	L	2 00 AM
B		3 00 AM
C		4 15 AM
D		5 30 AM
E		6 35 AM
F		7 45 AM
G		8 50 AM
H	A	10 00 AM

Do the schedules correspond in the seven requirements?

Yes.

What is the last No. 7 that can leave A on the old time-table?

No. 7 of Saturday.

What is the first No. 7 of the new time-table?

No. 7 of Monday.

If an opposing inferior train leaves H Saturday after 6:00 p.m. with an order to meet No. 7 at B, and it is at F at 2:30 a.m., Sunday, how must it be governed with respect to No. 7 of the new time-table?

Would not consider No. 7 until No. 7 of Monday becomes due, then clear its schedule as required by rule.

NEW TIME-TABLE EFFECTIVE 1 40 A M SUNDAY

OLD		
WESTWARD		
	1st CLASS	
	9	
	DAILY	
A		
B	L	1 10 AM
C		1 20 AM
D		1 45 AM
E		2 00 AM
F		2 10 AM
G		2 40 AM
H	A	3 10 AM

NEW		
WESTWARD		
	2nd CLASS	
	9	
	DAILY	
A	L	1 10 AM
B		1 30 AM
C		1 40 AM
D		2 05 AM
E		2 20 AM
F		2 30 AM
G	A	3 00 AM
H		

Do the schedules correspond in the seven requirements?

No.

In what do they fail to correspond?

In Class, Initial and Terminal stations.

If they fail to correspond in any one of the seven requirements, would it be equivalent to failing in all?

Yes.

What becomes of No. 9 of the old time-table at the time of change (1:40 a.m. Sunday)?

It ceases to exist.

When does No. 9 of the new time-table exist out of A for the first time?

1:10 a.m. Monday.

NEW TIME-TABLE EFFECTIVE 12 00 NOON TODAY

	OLD	
	WESTWARD	
		1st CLASS
		11
		DAILY
	A	L 8 10 AM
	B	8 30 AM
3rd	C	8 55 AM
2nd	D	9 20 AM
1st	E	9 40 AM
	F	10 00 AM
	G	10 20 AM
	H	A 10 30 AM

NEW		
WESTWARD		
	1st CLASS	
	11	
	DAILY	
A	L 2 00 PM	
B	2 20 PM	
C	2 45 PM	3rd
D	3 10 PM	2nd
E	3 30 PM	1st
F	3 50 PM	
G	4 10 PM	
H	A 4 20 PM	

Do the schedules correspond in the seven requirements?

Yes.

Is there a train authorized on the old time-table at 12:00 Noon, the time of change?

Yes, provided No. 11 of the old time-table has not been fulfilled over the entire run before 12 Noon.

If No. 11 of the old time-table is at an intermediate station at 12:00 Noon, is there a train authorized when the new time-table takes effect?

Yes. If No. 11 is at an intermediate station at 12:00 Noon, No. 11's schedule is only fulfilled up to such station at the time of change, the remaining portion being overdue, unfulfilled and not more than 12 hours late at 12:00 Noon, authorizes a train.

If called to leave A on No. 11 today at 2:00 p.m., on the new time-table, what should you first ascertain?

Whether No. 11 left A under the old time-table.

If No. 11 had left A before 12:00 Noon on the old time-table, may another No. 11 be run out of A at 2:00 p.m. on the new time-table? Why?

No. There can be only one schedule of the same number and day in effect.

If No. 11, which is due to leave A at 8:10 a.m., receives an order to run 2 hours late A to H, what time could it leave A and arrive at F and when could it leave F?

Could leave A 10:10 a.m., or any intermediate station B to E not less than two hours late on the old time-table and arrive at F as soon after leaving E as the run can be made under the rules, but not later than 12:00 Noon; and could not leave F until 5:50 p.m.

If No. 11 receives an order to run 2 hours late A to F, what time could No. 11 leave F on the new time-table?

At 3:50 p.m.

If No. 11 of today is annulled A to H, could No. 11 on the new time-table be run after 12:00 Noon? Why?

No, as No. 11 of the old time-table and new time-table correspond in the seven requirements, they are one and the same schedule and the annulment applies to No. 11 of today, as printed on both the old and new time-tables.

If No. 11 is fulfilled on the old time-table A to H before 12:00 Noon, could there be another No. 11 of the same date on the new time-table?

No.

If No. 11 makes E on the old time-table, what effect would this have on No. 11's time of the new time-table between A and E?

It could not be used. The schedule of No. 11 on the new time-table could not be again fulfilled A to E after 12:00 Noon.

If No. 11 of the old time-table leaves E before 12:00 Noon, would it nullify more of the schedule of No. 11 of the new time-table ?

Yes.

If No. 11 of the old time-table is delayed at E until 12:00 Noon or later, when and how may it proceed?

It may leave E at 3:30 p.m. on the new time-table, or within 12 hours thereafter.

If No. 11 is delayed at E until 12:00 Noon, of what value is the remaining portion of its schedule on the old time-table between E and H?

None.

If there were several sections of No. 11 at E at 12:00 Noon, the time of change, how should they be governed?

All sections of No. 11 could proceed from E at 3:30 p.m. as authorized by the rules, or within 12 hours thereafter.

If there are three sections of No. 11 out of A under the old time-table, the first section at E, second at D and the third at C at 12:00 Noon, the time of change, how should they be governed?

Third 11 could leave C at 2:45 p.m.; Second 11 could leave D at 3:10 p.m., and First 11 could leave E at 3:30 p.m., running far enough apart to comply with the rules relative to spacing trains.

If the third, or last section of No. 11 is at C at 12:00 Noon, what part of the schedule of No. 11 of the new time-table of today cannot be used?

That portion from A to, and including the arriving time at C.

If there are two sections of No. 11 and the first section reaches H on the old time-table before 12:00 Noon, may a first section run on the new time-table after 12:00 Noon?

No.

If the second, or last section does not leave A before 12:00 Noon, when and how may it leave on the new time-table?

At 2:00 p.m., as second, or last section of No. 11.

If Second 11 does not leave A before 12:00 Noon, could additional sections be authorized to follow Second 11 from A after 2:00 p.m.?

Yes.

If an opposing inferior train, ready to leave H at 11:00

a.m. is unable to get orders against the second section after checking the register and finding that First 11 only has arrived, how long must it remain at H for the second, or last section?

Until 12:00 Noon; then proceed, clearing Second 11 under the new time-table. If second, or last section of No. 11 does not arrive at H before 12:00 Noon, it must then respect the schedule of No. 11 as shown on the new time-table.

If there are two sections of No. 11, the first one reaches H before 12:00 Noon today and the second is at E at 12:00 Noon, how must the second be governed?

Wait at E until 3:30 p.m., then run as Second 11 on the new time-table E to H.

If the second (last section) is at E at 12:00 Noon, what effect would it have on the schedule of No. 11 of the new time-table A to E?

The schedule is fulfilled A to E.

If an opposing inferior train is at H at 12:00 Noon, and only the first section of No. 11 has arrived and it leaves H against Second 11 on the new time-table and makes E by 1:30 p.m., and finds Second 11 at E without signals, how should it be governed?

Proceed, as No. 11's schedule has been fulfilled A to E.

If an opposing inferior train leaves H at 12:00 Noon, before the arrival of Second 11, and fails to find Second 11 on the road on account of it waiting for time on an obscure track which is not in view of the main track and failing to notify opposing inferior trains, what effect would it have on the inferior train if it could not make A by 2:00 p.m.?

It would have to clear the time of No. 11 of the new time-table. Having met Second 11 without knowledge of same, the opposing inferior train could do nothing but clear the time of No. 11.

If the opposing inferior train meets Second 11 on the road with green signals waiting for time, how should it be governed?

Clear the time of No. 11 of the new time-table and wait for the last section.

NEW TIME-TABLE EFFECTIVE 12 01 A M SUNDAY

OLD	
WESTWARD	
	2nd CLASS
	13
	DAILY
A	10 00 AM
B	12 05 PM
C	2 00 PM
D	4 00 PM
E	5 00 PM
F	6 00 PM
G	6 30 PM
H	7 00 PM

Saturday's 13 Arrive D 12 01 AM

NEW	
WESTWARD	
	2nd CLASS
	13
	DAILY EXCEPT SUNDAY
A	L 11 00 AM
B	1 00 PM
C	3 00 PM
D	5 00 PM
E	6 00 PM
F	7 00 PM
G	7 30 PM
H	A 8 00 PM

Saturday's 13 on New Time-Table Leave D 12 15 AM

12 Hours Late 8 00 AM Sunday

Do the schedules correspond in the seven requirements on Sunday?

No.

Do they correspond in the seven requirements other days in the week?

Yes.

What is the date of No. 13 leaving A at 10:00 a.m. Saturday?

No. 13 of Saturday.

If No. 13 of Saturday is at D at 12:01 a.m. Sunday, how late is it on the old time-table?

Eight hours and one minute late.

If No. 13 of the old time-table is eight hours and one minute late at D at the time the new time-table takes effect and the schedule has not been fulfilled beyond that point, is there an authorized No. 13 of Saturday at D at the time of change?

Yes.

If there is a No. 13 of Saturday at D at 12:01 a.m. Sunday, and ready to leave at 12:15 a.m., how may it proceed?

It may proceed from D as No. 13 of Saturday of the new time-table, seven hours and fifteen minutes late.

What time has No. 13 of Saturday on the new time-table

to make H within the 12 hour limit?

Has until 8:00 a.m. Sunday.

If No. 13 of Saturday had not left A before 12:01 a.m. Sunday, could No. 13 of Saturday leave A on the new time-table at or after 12:01 a.m. Sunday?

No! No. 13 of the old time-table had been in effect at A, becoming 12 hours late before the new time-table became effective; and there could not be two No. 13's of Saturday in effect at A.

If No. 13 of Saturday has not been fulfilled out of B before 12:01 a.m. Sunday, could a No. 13 of Saturday run out of B on the new time-table at 12:01 a.m., Sunday.

Yes. No. 13's schedule of Saturday of the old time-table is only 11 hours and 56 minutes late at B at 12:01 a.m., Sunday, the time of change; therefore, we have a train authorized at 12:01 a.m., and No. 13 of Saturday, 11 hours and 1 minute late, at 12:01 a.m., could leave B any time before 1:00 a.m. Sunday.

If an opposing inferior train is ready to leave H any time before 8:00 am. Sunday and No. 13 of Saturday has not registered, could the inferior train proceed against No. 13 without orders?

No, as No. 13 of Saturday has until 8:00 a.m. Sunday to reach H.

NEW TIME-TABLE EFFECTIVE 12 01 A M SUNDAY

OLD	
WESTWARD	
	2nd CLASS
	13
	DAILY
A	L 10 00 AM
B	12 05 PM
C	2 00 PM
D	4 00 PM
E	5 00 PM
F	6 00 PM
G	6 30 PM
H	A 7 00 PM

NEW	
WESTWARD	
	2nd CLASS
	13
	DAILY EXCEPT SATURDAY
A	L 11 00 AM
B	1 00 PM
C	3 00 PM
D	5 00 PM
E	6 00 PM
F	7 00 PM
G	7 30 PM
H	A 8 00 PM

What days of the week do these schedules correspond in the seven requirements?

Every day except Saturday.

If No. 13 of Saturday of the old time-table is at D at 12:01 a.m., Sunday, how may it proceed? Why?

Proceed only as authorized by train order as there is no schedule on the new time-table to assume.

Could No. 13 of Saturday wait at D until 5:00 p.m Sunday and proceed?

No. A train of one day cannot assume the schedule of another day, even though they are the same number.

NEW TIME-TABLE EFFECTIVE 7 00 P M WEDNESDAY

OLD			NEW		
WESTWARD			WESTWARD		
	1st CLASS			1st CLASS	
	15			15	
	DAILY			DAILY	
A	L	2 00 AM	A	L	8 00 PM
B		2 30 AM	B		8 30 PM
C		3 00 AM	C		9 00 PM
D		3 30 AM	D		9 30 PM
E		4 00 AM	E		10 00 PM
F		4 30 AM	F		10 30 PM
G		5 00 AM	G		11 00 PM
H	A	5 30 AM	H	A	11 30 PM

Do the schedules correspond in the seven requirements?

Yes.

If No. 15's schedule is fulfilled on the old time-table Wednesday, could No. 15 run on the new time-table, leaving A at 8-00 p.m. Wednesday?

No. If there has been one No. 15 in effect Wednesday, there could not be another.

If No. 15's schedule of the old time-table was not fulfilled

Wednesday, could No. 15 run on the new time-table, leaving A at 8:00 p.m. Wednesday?

No. No. 15 of Wednesday having been in effect under the old-time table, even though no train ran on the schedule, it became 12 hours late over the entire run A to H, one hour and thirty minutes before the new time-table took effect; and as there has been one No. 15 in effect Wednesday, there cannot be another.

When does No. 15 of the new time-table exist for the first time out of A?

8:00 p.m. Thursday.

If an opposing inferior train is ready to leave H at 11:30 p.m., Wednesday, 4 hours and 30 minutes after the new time-table takes effect, and No. 15 of Wednesday does not appear on the register at H, could it leave H at 11:30 p.m.?

Yes. At 5:30 p.m. Wednesday, No. 15's schedule of Wednesday had been in effect and become twelve hours late over the entire run.

NEW TIME-TABLE EFFECTIVE 7 00 P M WEDNESDAY

OLD	
WESTWARD	
	1st CLASS
	15
	DAILY
A	L 5 00 AM
B	5 30 AM
C	6 00 AM
D	6 30 AM
E	7 05 AM
F	7 30 AM
G	8 00 AM
H	8 30 AM

15 at E at 7 00 PM

NEW	
WESTWARD	
	1st CLASS
	15
	DAILY
A	L 8 00 PM
B	8 30 PM
C	9 00 PM
D	9 30 PM
E	10 00 PM
F	11 00 PM
G	11 15 PM
H	A 11 30 PM

15 Leave E at 10 00 PM

If an opposing inferior train is ready to leave H at 11:30 p.m. Wednesday and No. 15 of Wednesday does not appear

on the register, should it leave H at 11:30 p.m. against the figures as shown on the new time-table at H? Why?

No, as No. 15 of Wednesday of the old time-table had not been in effect, or fulfilled, over the entire run prior to the time the new time-table became effective at 7:00 p.m., the schedule of No. 15 of the new time-table from E to H is in effect, No. 15's schedule of the old time-table being but 11 hours and 55 minutes late at E at time of change, authorizing a train E to H.

Would it be possible to start a No. 15 on the new time-table Wednesday from A, B, C, or D? Why?

No. No. 15 had been in effect at A, B, C and D, having become twelve hours late under the old time-table before the new time-table became effective at 7:00 p.m. Wednesday.

When does No. 15 of Wednesday lose both right and schedule under the new time-table between E and H?

When No. 15 of Wednesday is twelve hours late at these stations on the new time-table.

NEW TIME-TABLE EFFECTIVE 9 00 A M SUNDAY

OLD		
WESTWARD		
		1st CLASS
		17
		DAILY
A	L	11 00 AM
B		12 01 PM
C		1 00 PM
D		2 00 PM
E		3 00 PM
F		4 00 PM
G		5 00 PM
H	A	6 00 PM

NEW		
WESTWARD		
		1st CLASS
		17
		DAILY
A	L	8 00 AM
B		9 00 AM
C		10 00 AM
D		11 00 AM
E		12 01 PM
F		1 00 PM
G		2 00 PM
H	A	3 00 PM

Do the schedules correspond in the seven requirements?
Yes.

May a No. 17 be run on Sunday, the day of change? Why?
No. Although alike in the seven requirements, there is

not a train authorized on the old time-table at the time of change, to assume the schedule of the new time-table. No. 17 of the new time-table does not exist out of A until 8:00 a.m. Monday.

What was the last No. 17 authorized out of A on the old time-table?

No. 17 of Saturday.

If No. 17 was due to leave A on the old time-table at 9:01 a.m., would a No. 17 be authorized at time of change?

No.

If the new time-table should take effect at 11:30 a.m., Sunday, instead of 9:00 a.m., would there be a No. 17 authorized on the old time-table at the time of change?

Yes.

NEW TIME-TABLE EFFECTIVE 9 00 A M SUNDAY

OLD		
WESTWARD		
	1st CLASS	
	17	
	DAILY	
A	L	8 59 AM
B		9 25 AM
C		9 50 AM
D		10 10 AM
E		10 40 AM
F		11 10 AM
G		11 40 AM
H	A	12 10 PM

NEW		
WESTWARD		
	1st CLASS	
	17	
	DAILY	
A	L	9 01 AM
B		9 30 AM
C		9 45 AM
D		10 15 AM
E		10 45 AM
F		11 15 AM
G		11 45 AM
H	A	12 15 PM

Do the schedules correspond in the seven requirements?

Yes.

Is there a No. 17 authorized on the old time-table at 9:00 a.m. Sunday, the time of change? If so, how long has it been authorized?

Yes, one minute.

May No. 17 leave A at 9:01 a.m. on the new time-table,

Sunday, if it did not leave at 8:59 a.m. on the old time-table?

Yes.

NEW TIME-TABLE EFFECTIVE 12 01 A M SUNDAY

OLD		
WESTWARD		
	1st CLASS	
	19	
	Monday Wednesday Friday	
A	L	8 00 AM
B		8 55 AM
C		10 00 AM
D		11 00 AM
E		12 01 PM
F		1 00 PM
G		2 00 PM
H	A	2 30 PM

NEW		
WESTWARD		
	1st CLASS	
	19	
	Tuesday Thurs. Saturday	
A	L	8 00 AM
B		8 55 AM
C		10 00 AM
D		11 00 AM
E		12 01 PM
F		1 00 PM
G		2 00 PM
H	A	2 30 PM

Do the schedules correspond in the seven requirements?

No.

In what manner do they not correspond?

"Day of Leaving."

What is the last No. 19 that exists on the old time-table and the first on the new time-table?

8:00 a.m. Friday on the old time-table; and 8:00 a.m. Tuesday on the new time-table.

If the new time-table should take effect at 12:01 a.m. Saturday, when would the last No. 19 exist out of A on the old time-table, and the first on the new time-table?

8:00 a.m. Friday on the old time-table and 8:00 a.m. Saturday on the new time-table.

If the new time-table should take effect at 12:01 a.m. Friday, when would the last No. 19 exist out of A on the old time-table and the first on the new time-table?

8:00 a.m. Wednesday on the old time-table and 8:00 a.m. Saturday on the new time-table.

NEW TIME-TABLE EFFECTIVE 12 01 A M FRIDAY

OLD		
WESTWARD		
	1st CLASS	
	21	
	SATURDAY ONLY	
A	L	8 00 AM
B		8 55 AM
C		10 00 AM
D		11 00 AM
E		12 01 PM
F		1 00 PM
G		2 00 PM
H	A	2 30 PM

NEW		
WESTWARD		
	1st CLASS	
	21	
	SUNDAY ONLY	
A	L	8 20 AM
B		9 05 AM
C		10 20 AM
D		11 20 AM
E		12 20 PM
F		1 20 PM
G		2 20 PM
H	A	3 30 PM

Do the schedules correspond in the seven requirements?
No.

When does the last No. 21 exist out of A on the old time-table and the first on the new time-table?

8:00 a.m. on the old time-table, the last Saturday preceding Friday, the day of change; 8:20 a.m. on the new time-table the first Sunday following Friday, the day of change.

NEW TIME-TABLE EFFECTIVE 10 00 A M SUNDAY

OLD		
WESTWARD		
	1st CLASS	
	25	
	DAILY	
A	L	7 00 AM
B		7 30 AM
C		8 00 AM
D		8 30 AM
E		9 00 AM
F		9 30 AM
G		10 00 AM
H	A	10 30 AM

NEW	
WESTWARD	

As there is no schedule on the new time-table corresponding as required, what becomes of the schedule of No. 25 of the old time-table at 10:00 a.m., Sunday?

The schedule of No. 25 of Sunday ceases to exist after 10:00 a.m. If No. 25 is on the road, it can only proceed as may be authorized by train order, after 10:00 a.m.

If an opposing train holds a meet with No. 25 of Sunday, how should it be governed after 10:00 a.m.?

Proceed the same as if the order had not been issued.

NEW TIME-TABLE EFFECTIVE 12 00 NOON SUNDAY

OLD		
WESTWARD		
		1st CLASS
		27
		DAILY
A	L	10 00 AM
B		10 40 AM
C		11 20 AM
D		11 40 AM
E		12 00 NOON
F		12 40 PM
G		1 20 PM
H	A	2 00 PM

NEW		
WESTWARD		
		1st CLASS
		27
		DAILY
A	L	10 00 AM
B		10 40 AM
C		11 20 AM
I		11 40 AM
J		12 00 NOON
K		12 40 PM
G		1 20 PM
H	A	2 00 PM

Do the schedules correspond in the seven requirements?

No, they differ in route.

How long does the schedule of No. 27 of Sunday exist?

Until 12:00 Noon.

At 12:00 Noon has there been a No. 27 in effect under the old time-table whether a train has used it or not?

Yes.

What is the first time No. 27 can exist on the new time-table?

10:00 a.m. Monday

NEW TIME-TABLE EFFECTIVE 10 00 A M SUNDAY
SAME DIVISION OR SUBDIVISION

OLD		
WESTWARD		
	1st CLASS	
	29	
	DAILY	
A	L	8 10 AM
B		8 25 AM
C		8 40 AM
D		8 55 AM
E		9 15 AM
F		9 35 AM
G		9 56 AM
H	A	10 20 AM

NEW		
WESTWARD		
	1st CLASS	
	29	
	DAILY	
I	L	10 05 AM
J		10 15 AM
K		10 25 AM
D		10 35 AM
E		10 55 AM
F		11 15 AM
G		11 35 AM
H	A	11 50 AM

Do the schedules correspond in the seven requirements?

No they differ in initial stations and route.

How long does the schedule of No. 29 of Sunday exist?

Until 10:00 a.m.

At 10:00 a.m., has there been a No. 29 in effect on the old time-table, whether a train has used it or not?

Yes.

What is the first time No. 29 can exist on the new time-table and why?

At 10:05 a.m. Monday, as there has been one No. 29 in effect Sunday under the old time-table.

If the schedule has not been fulfilled on the old time-table between D and H prior to 10:00 a.m., Sunday, why cannot that portion of the new time-table be used Sunday?

As there has been one schedule in effect Sunday under the old time-table, the rule does not permit another of the same date.

NEW TIME-TABLE EFFECTIVE 12 01 A M SUNDAY
SUBDIVISION NO. 1

OLD	
WESTWARD	

NEW	
WESTWARD	
	1st CLASS
	55
	DAILY
A	L 10 00 PM
B	10 50 PM
C	11 30 PM
D	12 15 AM
E	1 00 AM
F	1 45 AM
G	2 20 AM
H	A 3 10 AM

SUBDIVISION NO. 2

OLD	
WESTWARD	

NEW	
WESTWARD	
	1st CLASS
	55
	DAILY
H	L 3 15 AM
I	4 00 AM
J	4 45 AM
K	5 30 AM
L	6 20 AM
M	7 00 AM
N	7 30 AM
O	A 8 00 AM

How many subdivisions are shown on the new time-table?
Two.

Is it understood that there is not a No. 55 on the old timetable corresponding in the seven requirements?

Yes.

When does No. 55 exist at A for the first time?

10:00 p.m. Sunday, the day of change.

When is No. 55 first due to leave H on Subdivision No. 2?

At 3:15 a.m. Sunday, the day of change.

Why does No. 55 exist out of H at 3:15 a.m. Sunday, 18 hours and 45 minutes before it exists out of A?

Subdivision No. 2 must be considered a separate railroad from Subdivision No. 1, as far as the movement of trains is concerned; and No. 55 may leave its initial station H on Subdivision No. 2 at its first leaving time after the new timetable takes effect.

If there were no subdivision, or if it were one subdivision from A to O, instead of two, would it be one schedule A to O, and what would be the initial and terminal stations?

Yes, one schedule A to O; A the initial and O the terminal stations.

If it were one division, or subdivision, A to O, when would No. 55 exist out of A for the first time?

10:00 p.m. Sunday, the day of change.

If it were one division, or subdivision, A to O, when would No. 55 exist out of H for the first time?

3:15 a.m. Monday, five minutes after it is due to arrive at H; as H is then an intermediate station.

How many times are given for a train at any station?

Not more than two.

Where one is given, what is it, unless otherwise indicated?

The leaving time.

Where two are given, what are they?

The arriving and the leaving time.

Where does time apply?

At the switch where an opposing train clears.

Where there is no switch, where does it apply?

At the station.

How are schedule meeting or passing stations indicated?

By figures in full-faced type.

When are both the arriving and leaving time of a train in full-faced type?

When both are meeting or passing times, or when one or more trains are to meet or pass it between those times.

Where there are one or more trains to meet or pass a train at any station, how is attention called to it?

By the numbers of the trains to be met or passed in small figures adjoining.

What do the following letters indicate when placed before the figures of the schedule:

"L"?

Leave.

"A"?

Arrive.

"s"?

Regular stop.

"f"?

Flag stop to receive or discharge traffic.

What must employes do whose duties may require them to give signals?

They must provide themselves with proper appliances, keep them in good order and ready for immediate use.

When must flags of the prescribed color be used?

By day, from sunrise to sunset.

When must lights of the prescribed color be used?

By night, and when day signals cannot be plainly seen.

Must signals be used in accordance with the rules?

Yes.

Must trainmen, yardmen, enginemen and all concerned keep a constant lookout for signals?

Yes.

How must those giving signals locate themselves?

So as to be plainly seen.

Are those giving signals required to give them so as to be plainly understood?

Yes.

Must care be exercised to avoid taking signals that may be intended for another train or engine.

Yes.

When backing or shoving a train, engine or cars, how must the disappearance from view of employe or light by which signals are given be construed?

As a stop signal.

What do the following color signals indicate:

Red?

Stop.

Yellow?

Proceed at reduced speed, and for other uses prescribed by the rules.

Green?

Proceed, and for other uses prescribed by the Rules.

Green and white?

Flag stop at the stations indicated on a train's schedule.

Blue?

Workmen are under or about engine, car or train.

Purple?

Stop. (Indication for dwarf signals.)

Who is authorized to remove a blue signal placed on car, engine or train?

Only the workman who placed it there.

Should cars be placed on a track in such a manner as to obscure a blue signal?

No.

What is the indication of hand, flag and lamp signals when used in the following manner:

Swung across the track?

Stop.

Held horizontally at arm's length?

Reduce speed.

Raised and lowered vertically?

Proceed.

Swung vertically in a circle at half arm's length across track?

Back?

Swung horizontally above the head, when standing?

Apply air brakes.

Held at arm's length above the head, when standing?

Release air brakes.

What signal is any object waved violently by anyone on or near the track?

Stop.

Must a fusee burning red on or near the track of an approaching train be extinguished?

Yes.

How may train then proceed?

At reduced speed.

If displayed beyond the first rail of an adjoining main track, will a fusee burning red apply to track on which train is running?

No.

Should fusees be so placed that fire will not be communicated to structures or cars of a following train?

Yes. If possible they should be thrown just to right of track in direction of movement, on engineman's side.

Is the use of fusees confined to night time?

No, they must be used by day when conditions warrant.

Should operators keep fusees on hand for emergency use when other stop signals fail?

Yes.

What is the indication of engine whistle signal when sounded as follows:

One short?

Apply brakes. Stop.

Two long?

Release brakes. Proceed.

One long, three short?

Flagman protect rear of train.

Four long?

Flagman may return from west or south, as prescribed by Rule 99.

Five long?

Flagman may return from east or north, as prescribed by Rule 99.

Two short?

Answer to 14 (k) or any signal not otherwise provided for.

Three short, when standing?
Back. Answer to 12 (d) and 16 (c).

Three short, when running?
Answer to 16 (d).

Four short?
Call for signals.

One long two short? (SINGLE TRACK)
To call attention of engine and train crews of trains of the same class, inferior trains and yard engines, and of trains at train-order meeting points, to signals displayed for a following section.

If not answered by a train, what action must be taken by the train displaying the signals?
Stop and ascertain the cause.

One long two short? (TWO OR MORE TRACKS)
To call attention of engine and train crews of trains of the same class and to inferior trains moving in the same direction and to yard engines, to signals displayed for a following section.

Two long, one short, one long?
Approaching public crossing at grade. To be prolonged or repeated until crossing is reached.

One long?
Approaching stations, junctions, railroad crossings at grade and . . . ., as may be required.

Two long one short?
Approaching meeting or waiting points. *See Rule S*-90.

One short, one long?
Inspect train line for leak or for brakes sticking.

Succession of short sounds?
Alarm for persons or live stock on the track.

What is the signal for train parted?
Three long blasts of the engine whistle.

How should train-parted signal be answered?
By trainman swinging arm, or lamp in circle at full arm's length.

Note: Train-parted signals have been eliminated from the Standard Code but may still be found in some books of rules.

When is it proper to sound four short blasts of engine whistle?

When signals are required, or when a signal is not understood, or when there is a failure to give them.

Should a train carrying green signals sound one long and two short blasts of the whistle when passing inferior trains, as well as trains that are met?

Yes.

Should the signal be given at both ends of the train met or passed?

Yes.

On double track, when one train passes another, both displaying signals, should both use the signal?

No, only the one passing. (A.A.R. Ruling Oct. 17, 1910.)

Should an engineman fail to whistle green signals, what must the inferior train do, having observed the signals?

The fact should be reported.

How should engineman of inferior train acknowledge whistle of green signals by superior train?

By two short blasts of engine whistle before the engine of superior train has passed.

If engineman fails to sound two long and one short blast of engine whistle at least one mile before reaching a meeting or waiting point, what action should conductor take?

Such action as may be necessary to insure safety.

On trains of passenger equipment, should conductor signal engineman first by air whistle approaching a meeting or waiting point to be answered by 14(n)?

This is not required by the Standard Code; however, a good many roads do require that conductor sound one short blast of communicating whistle to be answered by two long and one short blast of engine whistle.

Should enginemen use care in whistling and make their signals accurately?

Yes. If whistle signals are not made accurately, misunderstanding may arise.

Must the last blast of the crossing whistle be prolonged until the crossing is actually occupied?

Yes.

What signal does the explosion of two torpedoes indicate?
Proceed at reduced speed.

What does the explosion of one torpedo indicate?
The same as two, but the use of two is required.

Note: On some roads such as the Southern Pacific, the explosion of one torpedo is a Stop signal.

Where must torpedoes not be placed?
At stations or on public crossings.

What is the indication of communicating signal when sounded as follows:

Two short, when standing?
Start.
Two short, when running?
Stop at once.
Three short, when standing?
Back.
Three short, when running?
Stop at next passenger station.
Four short, when standing?
Apply or release air brakes.
Four short, when running?
Reduce speed.
Five short, when standing?
Recall flagman.
Five short, when running?
Increase speed.
Six short, when running?
Increase train heat.
One short, one long, one short?
Shut off train heat.
One long, when running?
Brakes sticking; look back for hand signals.
(*See the Standard Code note to Rule* 16.)
What will be displayed to the front of every train by night?
The headlight.
When must it be concealed or extinguished?
When a train turns out to meet another and has stopped clear of main track.

When must it be dimmed?

While passing through yards where yard engines are employed; approaching stations at which stops are to be made or where trains are receiving or discharging passengers; approaching train-order signals, junctions, terminals, meeting points, or while standing on main track, and on two or more tracks when approaching train in the opposite direction.

When an engine is running backward, what must be displayed by night on the rear of the tender?

A white light.

Before dimming the headlight, what must be observed?

It must be known that the main track switch next in front of the train is properly set for the opposing movement.

May the headlight be concealed before train has stopped on a siding?

No. The train must come to a full stop before headlight is extinguished.

If in doubt as to rear of train being clear of main track, may headlight be extinguished?

No. It must be known positively that the rear of the train is clear of main track.

May trainmen assume that engineman will keep the headlight displayed until train is clear of main track, thus relieving them from protecting the rear of their train?

No. Flag protection must be afforded the rear of their train at all times.

When a train turns out to be passed by another train, must the headlight be concealed?

No.

Is an engineman required to keep his headlight displayed until following trains are also clear of the main track?

No. The headlight will be concealed as soon as engineman knows that his own train has cleared.

May following trains assume that a leading train will keep headlight displayed until following train has cleared main track?

No.

If unable to clear at a siding, would it be good judgment for

an engineman to pull in behind cars in such a manner that his headlight is obscured?

No. He should stand back far enough that headlight can be seen until his train is properly protected by flagman.

How long should the headlight be left extinguished?

Until the markers of the train to be met have passed.

What will yard engines display, and where, by night?

The headlight to the front and rear.

When not provided with a headlight at the rear, what must be displayed?

A white light.

Must yard engines display markers?

No.

What signals will be displayed and where, by every train, as markers, to indicate the rear of a train, by day?

Marker lamps not lighted, or yellow (or green) flags as markers, one on each side of the rear.

By night?

Lights as markers, showing yellow (or green) to the front, and side and red to the rear.

How must these signals be displayed by night when clear of main track?

Lights as markers, showing yellow (or green) to the front, side and to rear.

How must these signals be displayed by night running with the current of traffic on .... track where there are three or more tracks?

Lights as markers, showing yellow (or green) to rear on the side next to the .... track in the direction of current of traffic and red on the opposite side.

Mow must these signals be displayed by night running on any track against the current of traffic where there are three or more tracks?

Lights as markers, showing yellow (or green) to the rear, with a red light on the platform or the cupola.

On a road which uses the right-hand track, how must these signals be displayed by night running against the current of traffic on two or more tracks?

Lights as markers, showing yellow (or green) to front and side and yellow (or green) to the rear on the side next to the main track on which the current of traffic is in the direction the train is moving and red to the rear on the opposite side.

What is required when a train is not equipped to display the prescribed signals to indicate the rear of the train?

Display a red flag by day and a red light by night.

(See the Standard Code diagrams.)

Should a freight train arrive with caboose apparently complete except that it has no markers, what does this indicate?

That the train is not complete.

When a train has turned out and is clear on the siding to be passed by another train, how should the markers be turned?

So as to display green (or yellow), to front, side and rear.

When must markers be changed back to display red to rear?

Before train attempts to occupy main track.

Should an engine be coupled to the rear of a train, where must the markers be displayed?

On the rear of the engine so coupled.

In this case, may markers be displayed also on the caboose?

No. That would indicate two trains coupled.

May a helper engine behind the caboose cut off at the summit of a grade and follow the train to be helped without flagging, on the assumption that he is displaying the markers and no train can leave the station in advance against him until he arrives?

No.

Should the engine be cut off or break off behind the caboose, is the conductor of the train relieved from notifying opposing trains that his train is incomplete?

No.

May an engineman leave a part of his train on the siding with markers displayed and take the front portion to a siding in advance depending upon the markers to protect his return movement to his train?

No.

If a freight train finds a caboose on a siding displaying markers with a number of cars ahead of it but no engine attached, what would good judgment require?

Ascertain whether the front portion of the train has broken away.

When should markers be removed?

After a train has stopped clear of the main track at the end of its run.

May an operator report a train to the dispatcher as having arrived before the markers are observed?

Not unless it can be positively learned that the entire train has arrived.

What will all sections except the last display in the places provided for that purpose on the front of the engine?

Two green flags, and, in addition, two green lights by night.

(See the Standard Code diagrams.)

What will extra trains display in the places provided for that purpose on the front of the engine?

Two white flags, and, in addition, two white lights by night.

(See the Standard Code diagrams.)

When two or more engines are coupled, shall the leading engine display the signals as prescribed by Rules 20 and 21?

Yes.

When one flag or light is displayed where in Rules 20 and 21 two are prescribed, what will it indicate?

The same as two; but the proper display of all train signals is required.

By what means of communication must each car of a passenger train be connected with the engine?

A communicating signal appliance.

What does a blue signal, displayed at one or both ends of an engine, car or train indicate?

That workmen are under or about it.

When thus protected, may it be coupled to or moved?

No.

Who will display the blue signals?

Each class of workmen.

Who alone is authorized to remove them?

The same workmen.

What action must first be taken when other equipment is to

be placed on the same track so as to intercept the view of the blue signals?

Notify the workmen.

What action must be taken when emergency repair work is to be done under or about cars in a train and a blue signal is not available?

The engineman and fireman will be notified and protection must be given those engaged in making the repairs.

If a train is blue-flagged, may the engine which is to take the train be coupled into it?

No, not until the blue signal is removed by the person who put it there.

If a blue signal is not available, what is required?

Before attempting emergency repair work, engine crew must be notified and protection arranged.

If a workman instructs that a blue signal be removed for him, may this be done?

No.

May an engineman move his engine a foot or two for the purpose of oiling around while the blue signal is displayed on his engine?

No.

May switchmen switch a train which has a blue signal displayed either at front or rear?

No.

If instructed to pick up a car on line at night, and car is found protected by blue flag instead of a blue light, may car be moved?

No.

When must white oscillating headlights be displayed?

At night, and during the day when weather conditions impair visibility.

Should it be extinguished when the standard white headlight is dimmed or extinguished?

Yes.

When must the red oscillating headlight be displayed?

By day, or by night, when train is stopped suddenly, under circumstances in which adjacent tracks may be fouled, or when head-end protection is required.

Must red oscillating headlight be extinguished as soon as its protection is no longer needed?

Yes.

What signal is conveyed to an approaching train on the same or adjacent track by a headlight burning red?

Stop, and be governed by conditions.

What signal is conveyed by an oscillating red light displayed to rear of train?

It is a signal for following trains on the same track to stop, and for trains moving in the same direction on adjacent track to move at restricted speed.

When must the oscillating red rear end light be displayed?

When train is stopped on or fouling main track, or when train is moving under circumstances in which it may be overtaken.

Do oscillating red lights at either end of train relieve crew from protecting their train in the manner prescribed by Rule 99?

No.

Note: Oscillating lights are not covered by Standard Code rules.

How must a signal imperfectly displayed, or the absence of a signal at a place where a signal is usually shown, be regarded?

As the most restrictive indication that can be given by that signal.

What is the exception to this requirement?

When the day indication is plainly seen, or when sufficient lights in a position or color position light signal are displayed to determine indication of the signal, the day indication will govern.

What action must be taken by engine and train crews using a switch where the switch light is imperfectly displayed or absent?

They must, if practicable, correct or replace the light.

When and to whom must a signal imperfectly displayed, or the absence of a signal at a place where a signal is usually shown, be reported?

Promptly to the proper authority.

What signal will be used to stop a train only at the flag stations indicated on its schedule?

Green and white.

When a signal, except a fixed signal, is given to stop a train, how must it be acknowledged, unless otherwise provided?

As prescribed by Rule 14 (g) or (h).

When must the engine bell be rung?

When an engine is about to move and while approaching and passing public crossings at grade and at stations, through tunnels and snow sheds.

Is the unnecessary use of either the whistle or the bell prohibited?

Yes.

What signals must watchmen, stationed at public crossings at grade, use when necessary to stop trains?

Stop signals.

What will they use to stop highway traffic?

Prescribed signals.

Must all members of engine and train crews when practicable, communicate to each other by its name the indication of each signal affecting the movement of their train or engine?

Yes.

What day signals will be used by flagmen?

A red flag, torpedoes and fusees.

What night signals will be used by flagmen?

A red light, torpedoes and fusees.

On single track, how is one train superior to another?

By right, class or direction.

How is right conferred?

By train order.

How is class and direction conferred?

By time-table.

Is right superior to class or direction?

Yes.

Between what trains is direction superior?

Trains of the same class.

On two or more tracks, how is a train superior to another train?

By right or class.

What is the order of superiority of trains by classes?

Trains of the first class are superior to those of the second; trains of the second class are superior to those of the third; and so on.

On two or more tracks, is there any superiority by direction as between trains of the same class?

No.

Are extra trains inferior to regular trains?

Yes.

Does superiority exist between extra trains?

No.

For how many hours are time-table schedules, unless fulfilled, in effect?

Twelve hours after their time at each station.

When do regular trains lose both right and schedule?

When more than twelve hours behind either their schedule arriving or leaving time at any station.

If No. 15 is due to arrive at D at 1:00 a.m. and leave at 2:00 a.m., when has it lost both right and schedule?

If it fails to arrive at D by 1:00 p.m., or leave D by 2:00 p.m.

May No. 15 flag into D at 1:10 p.m. and leave as No. 15 if it can get out by 2:00 p.m.?

No.

Having lost right and schedule, how may a train proceed?

Only as authorized by train order.

Referring to above question, if No. 15 flags into D at 1:10 p.m. and can leave by 2:00 p.m., may the dispatcher authorize its further movement as No. 15 by train order?

Yes.

If the dispatcher authorizes the train to run as No. 15 from D, must he provide an entirely new set of orders for the train?

Yes, with the exception of slow orders and annulments. All other orders became void when the train lost the schedule of No. 15.

Should a train arrive exactly 12 hours late on its schedule, would it lose both right and schedule?

No.

If No. 15 arrives at D at 1:01 p.m. (twelve hours and one minute late), and finds the train order signal in proceed position, may it leave as No. 15 without orders?

No.

Assuming that No. 15's time at C is 2:15 a.m., when may an inferior train leave C?

Not before 2:15 p.m.

If the dispatcher annuls No. 15 from H to D, can a train be authorized to use that portion of the schedule D to A?

Yes, as the schedule remains in effect D to A.

If Extra 575 West is ready to leave D at 1:00 p.m., when may it leave D, assuming No. 15 is to be a first class train?

Not before 2:01 p.m., when No. 15 becomes more than twelve hours late on its leaving time.

On single track, what is required before a train may leave its initial station on any subdivision, or a junction, or pass from one of two or more tracks to single track?

It must be ascertained whether all trains due, which are superior, have arrived or left.

When a train is restricted for an opposing extra at a register station, what must the restricted train know before leaving?

That the extra has arrived. It must either see and identify the extra, or have a train-order check stating the extra has arrived; or have the restriction annulled.

On two or more tracks, what is required before a train may leave its initial station on any subdivision or a junction?

It must be ascertained whether all superior trains due have left.

How are stations at which train registers are located designated?

By time-table.

What is required in starting trains from a point where there is no train register, or when the conductor and engineman are to be relieved of the duty of ascertaining what trains have arrived or left?

A train-order check of trains.

If the initial station is on double track and the register is checked and found clear, is it necessary to stop at the end of double track and check again for the same trains?

Yes. Signals for following sections may be taken down, or displayed from, any register station.

CHECKING TRAIN REGISTER

NOTE:-ALWAYS USE TIME-TABLE IN CHECKING TRAIN REGISTER

SOUTHWARD				TIME-TABLE No. 24 EFFECTIVE 12:01 AM NOV. 3, 1940	NORTHWARD			
SECOND CLASS		FIRST CLASS			FIRST CLASS		SECOND CLASS	
67 DAILY	65 DAILY	3 DAILY	1 DAILY	STATIONS	2 DAILY	4 DAILY	66 DAILY	68 DAILY
8:00 AM	9:15 PM	8:30 PM	7:30 AM	A	12:50 PM	2:50 PM	4:30 AM	2:30 PM
8:40 AM	9:40 PM	8:45 PM	7:45 AM	K	12:10 PM	2:20 PM	3:30 AM	1:30 PM
10:20 AM	11:10 PM	9:55 PM	8:55 AM	R	11:25 AM	1:30 PM	2:30 AM	12:30 PM
10:45 AM	11:45 PM	10:30 PM	9:30 AM	Z	11:10 AM	1:00 PM	1:30 AM	10:50 AM
DAILY	DAILY	DAILY	DAILY		DAILY	DAILY	DAILY	DAILY

CHECK OF TRAIN REGISTER AT 11:15 AM SHOWS:

FIRST 1 HAS ARRIVED; GREEN SIGNALS.
NO. 3 HAS NOT ARRIVED.
NO. 65 HAS NOT ARRIVED.
NO. 67 HAS ARRIVED; NO SIGNALS.
FIRST 2 DEPARTED ON TIME; SECOND 2 HAS NOT LEFT.
NO. 4 NOT DUE TO LEAVE UNTIL 1:00 PM.
NO. 66 HAS DEPARTED.
NO. 68 HAS NOT DEPARTED.

EXTRA 798 NORTH READY TO LEAVE Z AT 11:15 AM.

May the dispatcher relieve conductors from stopping to check registers?

Yes, by furnishing a train-order check of trains.

When a train is detoured, what must it have before leaving an intermediate station at which it returns to its own rails?

A train order authorizing its movement.

If a train is given a schedule at its initial station and is later detoured between two intermediate points, may it assume that schedule upon returning to its own line?

Under A.A.R. Code Rules, yes. A good many roads have rules which do not permit this, requiring that the train be put back on its schedule by train order at the point where it returns to its own line.

Referring to chart on page 277, since First 1 has arrived, what does the Extra need to leave against the schedule of No. 1?

Needs an order on Second 1.

No. 3 is more than 12 hours late; what does the Extra need on No. 3?

Nothing.

No. 65 is more than 12 hours late at R but not at Z; what does the extra need on No. 65?

Needs an order on No. 65.

No. 67 has arrived; does the Extra need anything on No. 67?

No.

First 2 departed on time; Second 2 has not left; what does the Extra need?

Authority to run ahead of Second 2.

Outside automatic block signal territory, if First 2 did not leave until 11:10 a.m., what time could the Extra leave behind it?

11:20 a.m., or 10 minutes behind it.

Does the Extra need an order to run ahead of No. 4?

No.

No. 66 has left; does the Extra need anything on No. 66?

No.

No. 68 has not left; does the Extra need authority to run ahead of No. 68?

No. Although the rules of some roads require that the Extra know whether or not No. 68 has left, most of them make provisions for extra trains to run ahead of second-class trains.

Before a train leaves its initial station on any subdivision, what must be obtained?

A clearance card, unless otherwise especially provided for by time-table or rule.

Should No. 1 be due to leave at 12:01 a.m., the clearance card be dated the day before and timed at 11:45 p.m., together with an order dated several days before (as might be in case of slow-orders), may the conductor understand his train is cleared properly?

Yes.

What must be received before the engineer may start movement of a train in either direction?

A proper signal.

Except when proceeding on block or interlocking signal indication after train has been stopped by such signal, who will give the proceed signal to start a passenger train?

The conductor.

When practicable, what signal will the conductor use?

The communicating signal.

What must be received before a train is backed?

Proper signal from the rear end.

When required, must proper rear end protection be afforded before backup move or a passenger train is started?

Yes.

May trains of one schedule pass trains of another schedule of the same class?

Yes.

May second and inferior class and extra trains pass and run ahead of second and inferior class and extra trains?

Yes.

May a section of a schedule pass and run ahead of another section of the same schedule?

Yes, first exchanging orders, signals and section numbers.

From where will a change of sections be reported to the dispatcher?

From the first available point of communication.

When trains are running in sections, with whom does the responsibility rest for a following section passing a leading section of the same schedule, except as provided by the rules?

With the leading section, as well as the following section.

Unless otherwise provided, how must an inferior train clear a first-class train, or a train made superior by train order, in the same direction?

At the time the first-class train, or train of superior right, is due to leave the next station in the rear where time is shown.

Is there any exception to this rule?

Yes. The majority of roads require that where the running time of the superior train is less than ten minutes between stations, a ten-minute clearance will be observed. Other roads have cut this minimum down to five minutes.

In automatic block signal territory, what clearance is required?

The inferior train must clear the block sufficiently to avoid delay to the first-class train, or train of superior right.

Should an inferior train fail to clear as prescribed by Rule 86, what must be done?

Protect as prescribed by Rule 99.

What is required of an inferior train with respect to opposing superior trains?

It must keep out of the way.

When an inferior train fails to clear the main track by the time required by rule, what must be done.

Train must be protected as prescribed by Rule 99.

Must train pull into siding when practicable?

Yes.

If necessary to back in, or back out, what must first be done?

Train must first be protected as prescribed by Rule 99, unless otherwise provided.

At meeting points between trains of different classes, which train must take the siding; and how much must the inferior train clear the superior train?

The inferior train must take the siding and clear the superior train not less than five minutes.

What should be understood by "at least five minutes"?

The inferior train must be on the siding, clear of the main track, with switches locked for the main track, at least five minutes before the leaving time of the superior train.

If the inferior train arrives at the switch five minutes before the superior train is due to leave, would it comply with the rule?

No.

Should the superior train, holding the main track, set the switch for the inferior train to take siding, if it arrives first?

Yes.

Where must the superior train stop?

Clear of the switch to be used by the inferior train in clearing.

What is required of the engineman of each train before reaching a meeting or waiting point?

He must give signal 14 (n), two long and one short blasts of whistle, at least one mile before reaching the point of restriction.

Should this signal be given if the train is restricted by a wait order?

Yes, if the train is actually in advance of its wait.

Should a stop be made after signal 14(n) has been given, may a proceed signal be given before the train to be met has arrived, or authority is held to proceed against it?

No. The engineman should be verbally instructed as to any desired move.

How will extra trains be governed with respect to other extra trains, unless otherwise provided?

As directed by train orders.

Unless otherwise provided, at meeting points between extra trains, which train must take the siding.

The extra moving in the inferior time-table direction.

On single track, may extra trains be run without train orders?

No, except in territory where rules governing movement of trains and engines by block signals are in effect.

At a meeting point between an extra and a work extra, which train will take the siding?

The work extra, even though it be moving in the superior time-table direction. (A.A.R. decision.)

At schedule meeting point between trains of the same class, when must the inferior train be clear of the main track?

Before the leaving time of the superior train.

Must the superior train stop at schedule meeting points?

Yes, if the train to be met is of the same class unless the switch is properly lined and the track clear.

Unless some form of block system is used, how many minutes must trains in the same direction keep apart?

Not less than five minutes, except when closing up at stations.

CLEARING TRAINS

SOUTHWARD FIRST CLASS			NORTHWARD FIRST CLASS	
NO. 1	NO. 3	STATION	NO. 2	NO. 4
10:15 AM	8:01 AM	SOUTH A SIDING	10:25 AM	8:30 AM
10:20 AM	8:03 AM	B YARD T. O. SIG.	10:20 AM	8:25 AM
10:25 AM	8:14 AM	C SIDING NORTH	10:10 AM	8:14 AM

DISTANCES—Between A and B—2 Miles
Between B and C—5 Miles

How many minutes must a train following a train carrying passengers keep behind it?

Not less than ten.

Note: A good many roads require that all trains be spaced not less than ten minutes.

May a train leave a station in advance of its schedule leaving time?

No.

Within yard limits, how may the main track be used?

Without protecting against second and inferior class, extra trains and engines.

How must second and inferior class, extra trains and engines move within yard limits?

At yard speed, unless the main track is known to be clear.

In clearing first-class train at a yard, is it understood that the term "next station in the rear" applies to the next station in the direction of any approaching first-class train?

Yes.

At meeting points within yard limits, what clearances must be observed by inferior trains.

The same clearances required elsewhere. These are not altered by Rule 93.

Referring to chart on page 282, outside automatic block signal territory, by what time must a southward inferior train be in clear for No. 1 at B?

10:10 a.m.

Outside automatic block signal territory, by what time must a yard engine or other engines be in clear for No. 1 at B?

10:10 a.m.

Outside automatic block signal territory, by what time must a southward inferior train or a yard or other engine be in clear for No. 3 at B?

7:56 a.m.

If a northward inferior train met No. 1 at B, what time would the northward inferior train be in clear in siding?

By 10:15 a.m.

At what point does the time (10:20 a.m.) of No. 1 at B apply?

At the switch where the opposing train enters the siding, which in this case, if there is no track designated as a siding, the recognized northward entrance to the yard.

Outside automatic block signal territory, by what time must a yard or other engine be in clear at B for No. 2?

10:10 a.m.

If a southward inferior train met No. 2 at B, by what time must the southward inferior train be in clear in siding for No. 2?

10:15 a.m.

At what point does the time of No. 2 at B (10:20 a.m.) apply?

At the switch where the opposing train enters the siding, which in this case, since there is no designated siding, is the southward recognized entrance to the yard.

Outside automatic block signal territory, by what time must a northward inferior train, or a yard engine or other engine be in clear at B for No. 4?

8:14 a.m.

By what time must a southward inferior train which will meet No. 4 at B, be in clear in siding for No. 4?

8:20 a.m.

B is the schedule meeting point for Nos. 1 and 2. At what time must No. 1 be in clear at B for No. 2?

Before the leaving time of No. 2 (10:20 a.m.).

No. 3 does not make the schedule meeting point (C) for No. 4, only makes B, what time must No. 3 be in clear at B for No. 4?

8:20 a.m.

Note: The answers above as to clearing trains are based on rules which require a minimum five-minute clearance. The majority of roads require not less than ten minutes.

When a train overtakes another so disabled that it cannot proceed, will it pass if practicable, and if necessary assume the schedule and take the train orders of the disabled train, proceeding to the next available point of communication, where it will report to the dispatcher?

Yes.

If the delayed train is of greater importance than the train which overtakes it, and all it needs is an engine, should the train which overtakes the disabled train give it its engine and permit the disabled train to move first?

Yes.

May more than one exchange be made, if necessary?

Yes.

If two trains are directed to meet a train at C and one of them becomes disabled at B, should the conductor of the disabled train accompany the train which passes him at B, or may the train which is being delayed at C be gotten out by sending word to it by the train which passes the disabled train?

The conductor of the disabled train should accompany the train which passes it at B and return on the opposing train from C.

If there are trains waiting at blind sidings between the disabled train and the next available point of communication, what must the train do which exchanges with and passes the disabled train?

It must stop, identify itself and explain the situation.

Why is the direction given to report to the dispatcher at the next available point of communication?

To enable the dispatcher to arrange for the movement of the trains thereafter as he deems proper.

When able to proceed, what does the disabled train do?

It assumes the right or schedule and takes the train orders of the last train with which it has exchanged and proceeds to and reports from the next available point of communication.

May two or more sections be run on the same schedule?

Yes.

Has each section equal time-table authority?

Yes.

May a train display signals for a following section, except as prescribed by Rule 85, without orders from the train dispatcher?

No.

May signals be ordered displayed to, or taken down at, other than a register station for the train displaying signals?

No.

On roads where it is permissible to create a following section at an intermediate point not a register station, how is the movement protected?

A copy of the order creating the section must be given to all inferior trains, or trains of the same class in both directions, at or before reaching such point, until the following section has left.

If signals are displayed to an intermediate register station where trains do not stop regularly but are furnished a train-order check of trains, should the dispatcher use extra precautions?

Yes. Until such time as the following section has arrived, or the newly created section has left that point, all trains affected should be brought to a stop before delivery of the order, as their superiority is actually restricted at that point.

Unless otherwise provided, may extra trains be run on two or more tracks without train orders?

No.

What are the requirements with respect to the movement of work extras on double track?

They must move with the current of traffic unless otherwise directed.

When a train is authorized by special instruction in the time-table to run extra on double track without running orders, moving with the current of traffic, and a train order is received authorizing a section of double track to be used as a single track, or an extra is moved against the current of traffic, what must be obtained before using such section of track?

Train order authority to run extra.

How must trains approach the end of double track, junction, railroad crossings at grade and drawbridges?

They must approach prepared to stop, unless the switches and signals are right and the track is clear. Where required by law, trains must stop.

After stopping, what whistle signal must engineman sound before starting to cross?

Two long blasts.

When a train stops under circumstances in which it may be overtaken by another train, when must the flagman go back?

Immediately.

What must he take with him?

Flagman's signals.

What are flagman's signals: By day? By night?

By day, red flag, torpedoes and fusees. By night, red light, torpedoes and fusees.

How far back must he go?

A sufficient distance to insure full protection.

What signals must he place and display?

Place two torpedoes, and when necessary, in addition, display lighted fusees.

When may the flagman return?

When recalled by engine whistle signal and safety to the train will permit.

When will he leave the torpedoes and a lighted fusee?

When the conditions require.

By whom must the front of a train be protected in the same way when necessary?

By the forward trainman or fireman.

When a train is moving under circumstances in which it may be overtaken by another train, what action must the flagman take?

Such action as may be necessary to insure full protection.

What is required by night, or by day, when the view is obscured?

Lighted fusees must be thrown off at proper intervals.

If a train finds a burning fusee in ditch or at foot of embankment at the side of the track on which the train is moving, must it be respected?

Yes.

On two or more tracks, if a burning fusee is found between the two main tracks or between the rails of the opposite track, what will be done?

Train will stop, then proceed at restricted speed, but fusee will not be extinguished.

On two or more tracks, if a fusee which is thrown from a train bounces across the opposite track, should trainmen throw off another fusee to proper location?

Yes.

When day signals cannot be plainly seen, owing to weather

or other conditions, what signals must be used in addition to day signals?

Night signals.

Who are responsible for the protection of their trains?

Conductors and enginemen.

How would you determine what is a sufficient distance?

By taking into consideration curves, grades and other physical characteristics, weather conditions, maximum speed of trains and any other factors which may affect safety, bearing in mind that you must always go back the distance the rules require; namely, sufficient distance to insure full protection.

If flagman is recalled before having reached the required distance to give train full protection, how will he be governed?

Continue to go back until a sufficient distance is reached to insure full protection, placing two torpedoes and in addition by night or in bad weather, or at any time when necessary, leave a burning fusee.

Do you understand that block signal indications do not modify the requirements of Rule 99?

Yes.

When it becomes evident that rear of train requires protection, what must the engineman immediately do?

Sound one long, three short, Rule 14 (c), for flagman to protect rear of train.

If necessary, must the signal be repeated until protection is assured?

Yes.

When a flagman is sent out with specific instructions affecting the authority of a train to proceed, must such instructions be in writing?

Yes.

When a flagman is sent to a station on a train, where will he ride, and where will engineman stop to let him off?

Flagman must ride on engine, and engineman must stop and let him off at the first switch at the station to which he is sent.

When the flagman goes back to protect the rear of the train and is left behind, who will take his place on the train?

Another trainman.

What is required when any known condition not covered by the rules interferes with the safe passage of trains?

Trains must be fully protected.

What must be done when any employe observes a condition that may result in an accident or interruption of train operation?

He must make prompt report, and if necessary, arrange proper protection.

Must the telegraph or telephone be depended upon to notify other trains when a defect or condition is found which might cause an accident?

No, a flagman must be left.

If any member of a train or engine crew has reason to believe their train has passed over any dangerous defect, what must be done?

Train must be stopped at once, proper protection provided, and prompt report made to the train dispatcher.

When practicable, should report be made to the nearest track or bridge foreman?

Yes.

When indication of severe storm, high water, fire or any condition which threatens damage, how must trains proceed?

At restricted speed.

If in doubt as to being able to proceed safely, what must be done with train?

Place it on siding to remain there until it is safe to proceed.

Must conductors and enginemen make careful inquiries at all stopping places, and when advisable, make extra stops to ascertain the extent and severity of storms; examine bridges and culverts or other places subject to damage by high water, and if any indication of danger, make report to train dispatcher?

Yes.

Shall any train be permitted to pass over a damaged steel span of a bridge or trestle until examination has been made and it is known that structure is safe for traffic?

No.

When a train is disabled or stopped suddenly by an emergency application of the air brakes or other causes, what must

be done with reference to adjacent tracks as well as tracks of other railroads that are liable to be obstructed?

The situation must at once be protected until it is ascertained that the tracks are safe and clear for the movement of trains.

When for any reason an engine leaves its train or part of its train on the main track, must a sufficient number of hand brakes be set, when necessary, to keep train from moving?

Yes.

By night, or when the view is obscured, what signal must be placed on front end of rear portion of train?

A red light.

When safety requires, what additional protection must be afforded to warn engineman returning?

Torpedoes must be placed a safe distance in advance of rear portion of the train, and when conditions require, a flagman must protect returning engine.

On heavy grades, when stopping on main track or siding, when cutting engine off a train or cars at stations to do work, or at stops of unusual length, what must be done to hold a train or cars?

A sufficient number of hand brakes must be set and the air released.

When cars are pushed by an engine, except when shifting or making up trains in yards, and even then when conditions require, where must a trainman take a conspicuous position?

On the leading car.

When shifting over public crossings at grade, not protected by a watchman, or manually controlled crossing signals or gates and same are known to be functioning, who must protect the crossing from a point on the ground on the crossing?

A member of the crew.

Must all movements over the crossing be made only on his signal?

Yes.

Is such protection required when facing end of leading car is equipped with back-up hose and air whistle handled by trainman?

No.

Must cars standing on tracks be properly secured?

Yes.

Must they clear other tracks and street, highway or private crossings and must they be left as far back as practicable?

Yes.

When coupling or shoving cars, must proper precaution be taken to prevent damage or fouling other tracks and sufficient hand brakes set?

Yes.

In shoving yard tracks, must it be known there is sufficient room in the track to hold the cars?

Yes.

If necessary to shove the entire length of the track, what must be done?

The cars must be coupled and a man sent to head end to protect the movement.

When switching or placing cars, must employes know that they are not left standing so close as to not fully clear passing car on adjacent tracks, or cause injury to employes riding on side of cars?

Yes.

When necessary to control cars by hand brakes, when must it be known that such brakes are in good order?

Before cars are cut off.

When are running switches permitted?

When they can be made without danger to employes, equipment or contents of cars.

What must be known before making the running switch?

That track is clear and the switches and brakes are in working order.

Must the movement be carefully made and the engine run on straight track, when practicable?

Yes.

Should a proper understanding be had between members of crew before making running switch?

Yes.

May trains block a public crossing longer than five minutes when it can be avoided?

No.

When passenger trains are at stations, must freight trains be opened to allow passageway?

Yes.

May trains be permitted to stand across another railroad, when practicable to avoid it?

No.

Are conductors responsible for the position of switches used by them and their trainmen, except where switch tenders are stationed?

Yes.

Must switches be properly lined after having been used?

Yes.

May a switch be left open for a following train or engine unless in charge of a trainman of such train or engine, or an assigned switch tender?

No.

When practicable, who must see that switches near the engine are properly lined?

The engineman.

May train or engine foul a track until switches connected with the movement are properly lined?

No.

When a train or engine is waiting to cross from one track to another, and during the approach or passage of a train or engine on track involved, how must all switches connected with the movement be secured?

In the normal position.

When may switches be restored to normal position?

Not until the movement is completed, or clear of the main track involved.

What is necessary when trains or engines are required to be reported clear of main track?

Report must not be made until switch has been secured in its normal position.

When employes set switches for the main track, must they test the lock to know that it is secured, and see that the switch points fit properly?

Yes.

What will be done when main track switch locks are found defective or missing?

They must be replaced immediately, or the switch securely spiked for main track movements.

Must the fact be reported to the train dispatcher, and when practicable, to the track foreman?

Yes.

When a main track switch has been set for a train, what position will the person attending the switch take?

He must go to the opposite side of the main track, and not return to the operating switch stand until the movement has been completed.

If on two or more tracks, where must he stand?

When not practicable to go to opposite side of track, he must stand at least twenty feet from operating switch stand.

May an employe who is to attend a main track switch for train or engine in siding for purpose of meeting an opposing train go beyond the head end of train or engine in siding until opposing train has passed the main track switch?

No, not for the purpose of attending the switch.

When practicable, where must employe alighting from moving train to line main track switch to normal position get off?

The rear end of rear car.

On what side of train must he get off, and when may he cross over to switch stand?

On side of train opposite the operating switch stand, and he may not cross over until train is in clear.

On a light engine, must you get off on opposite side from switch stand?

Yes.

Must derails be kept set and locked in derailing position after having been used, whether or not there are cars on the tracks?

Yes.

How must trains or engines using a siding or yard track proceed?

At yard speed.

May sidings of an assigned direction be used in a reverse direction, unless authorized by the train dispatcher?

No.

In an emergency, how may it be done?

Under flag protection.

May cars be left on sidings when possible to avoid it?

No.

When a siding is obstructed, who must be notified by wire?

The train dispatcher.

Who are responsible for the safety of the train and the observance of the rules?

Both the conductor and the engineman.

Under conditions not provided for by the rules, must every precaution be taken for protection?

Yes.

How must trains or engines be run in passing a train receiving or discharging traffic at a station, except where proper safeguards are provided?

At restricted speed.

Should a train or engine pass between the train receiving or discharging traffic and the platform of the station?

No, unless the movement is properly protected.

In case of doubt or uncertainty, what course must be taken?

The safe course.

Who must observe passing trains for defects?

All employes, as far as practicable.

Must trainmen of freight and passenger trains, yardmen and operators observe passing trains for defects?

Yes.

Where will operators at intermediate stations stand when trains are passing, unless excused by train dispatcher?

On station platform.

What are the defects to be looked for?

Brakes sticking, wheels sliding, brake rigging down, swinging doors, hot journals, protruding objects, lading dangerously shifted, evidence of fire, or any other condition which will endanger movement of train.

What signals must employes noting such defects give?

Stop signals.

What further action must be taken when communication with the train dispatcher is possible?

Train dispatcher must be notified of such defects.

If nothing irregular is noted, what signals will employes observing trains for defects give to the rear of the passing train?

Proceed signals.

When must train and engine crews be on lookout for signals from employes observing trains for defects?

When passing other trains, interlockings, open train-order offices, coal, water or other stations, and points where maintenance-of-way men are working.

Must train and engine crews, when practicable, exchange signals with employes observing trains for defects?

Yes.

Where two main tracks are in service, must trains keep to the right, unless otherwise provided?

Yes.

Where three or more tracks are in service, how will their use be indicated?

By special instruction.

When a train crosses over to, or obstructs another track, unless otherwise provided, what must first be done?

Movement must be protected as prescribed by Rule 99 in both directions.

For movements not provided for by time-table, by whose authority and over whose signature will train orders be issued?

Usually the chief train dispatcher.

Must train orders only contain information or instructions essential to such movements?

Yes.

Must they be brief and clear, in the prescribed forms, when applicable; and without erasure, alteration or interlineation?

Yes.

May figures in train orders be surrounded by brackets, circles or other characters?

No.

Must each train order be given to all employes addressed in the same words?

Yes.

Must train orders, except restricted speed orders, be numbered consecutively each day beginning at midnight?

Yes.

How must train orders be addressed?

To those who are to execute them, naming the place at which each is to receive his copy.

How must those for a train be addressed?

To the conductor and engineman, and also to any one who acts as its pilot.

Must a copy for each employe addressed be supplied by the operator?

Yes.

Note: Some roads require that each engine crew be given two sets of orders and that, in addition, a set be provided for the flagman.

How must orders be addressed to operators, restricting the movement of trains, be respected by conductors and enginemen?

The same as if addressed to them.

Must enginemen and firemen, and, when practicable, forward trainmen, read train orders, check with each other and have a definite and proper understanding of their requirements?

Yes.

Must conductors and, when practicable, trainmen, read train orders, check with each other and have a definite and proper understanding of their requirements?

Yes.

What record must be made of each train order?

It must be written in full in a book provided for that purpose.

What must be recorded with it?

The names of those who have signed for the order; the time and the signals which show when and from what office the order was repeated and responses transmitted; and the train dispatcher's initials.

When must these records be made?

At once, and never from memory or memoranda.

How will regular trains be designated in train orders?

As "No. 10," and sections as "Second 10," adding engine numbers, unless otherwise provided.

How will extra trains be designated?

By engine numbers, and the direction as "Extra 798, 'North' or 'South'."

How will an engine of another company be designated?

The initials as well as the engine number will be used, as "Extra MKT 868 North."

May even hours as "10 00 A.M." be used in stating time in train orders?

No.

In transmitting train orders by telegraph, may time be stated in figures only?

Yes.

Do you understand that figures may be duplicated in words when means of communication is not clear and distinct or when in the judgment of the dispatcher it is necessary?

Yes.

In transmitting and repeating train orders by telephone, how may numbers of trains and engines in the address be stated?

Pronounced and then spelled, letter by letter.

How *must* all stations and numerals in the body of an order be stated?

They must first be plainly pronounced and then spelled, letter by letter, thus: Aurora, A-u-r-o-r-a, and One Naught Five, O-n-e N-a-u-g-h-t F-i-v-e.

Will the letters duplicating names of stations and numerals be written in the train-order book, or upon train orders?

No. (Some roads require that time be spelled out in orders.)

When train orders are transmitted by telegraph, how will the train dispatcher write, check and underscore?

Write and check the order from the first repetition, and underscore each word and figure at the time of each succeeding repetition.

When transmitting by telephone, when must train dispatcher write and underscore?

He must write the order as he transmits it and underscore each word and figure at the time of each repetition.

When two or more engines are coupled, shall the number of each engine be used in train orders?

Some roads require this and, in addition, use the word coupled as Extra 798 and 799 coupled East. The majority of roads require that only the number of the leading engine shall be used in train orders.

When two extras are mentioned in train orders, shall the word "two" be used?

Yes. For example, "meet two Extras 798 and 799 North."

What signal will be used by the train dispatcher to transmit train orders?

The signal "31" or the signal "19" followed by the direction and number of copies must be given to each office addressed. Where roads use only a single order form, the "31" or "19" is omitted.

When a train order is sent to two or more offices, how must it be transmitted?

Simultaneously to as many of them as practicable.

When not sent simultaneously to all, to which train must the order be sent first?

To the superior train.

Must the several addresses be in the order of superiority of trains, each office taking its proper address?

Yes.

When practicable, except in automatic block signal territory, who must be included in the several addresses?

The operator at the meeting or waiting point.

To whom must operator deliver copies of such orders?

To trains affected until all have arrived from one direction.

May a train order be sent to a train at the meeting or waiting point if it can be avoided?

No.

When an order is so sent, what is required?

The fact will be stated in the order and special precautions must be taken to insure safety.

How must a train that is advanced to a meeting or waiting point, where the opposing train receives the order, approach such station?

With caution.

If the train-order signal is located so that there is danger of opposing trains overrunning siding switch, what must be done?

A flagman must protect ahead as prescribed by Rule 99.

How and when must operators receiving train orders write them?

In manifold during transmission.

Must operators retain a copy of each train order?

Yes.

What train orders may be typewritten?

Some roads permit all orders to be typewritten; other permit only slow orders to be typewritten.

What part of all orders must be in the operator's handwriting?

The time, complete, and the last name of the operator.

If the requisite number of copies cannot be made at one writing, how must operators make others?

From the original copy.

Must they repeat to the train dispatcher from the new copies each time additional copies are made?

Yes.

When orders are recopied, will the original date of issue, time completed, and operator's name be changed?

No.

When an error is made in transmitting a train order, and before it has been repeated or the X response sent, what must be done?

All copies must be immediately destroyed, the order marked "Void" in train-order book, and if reissued given another number.

If an order in which an error has been made has been repeated, or the X response sent, what must be done?

It must be annulled.

When a "31" order has been transmitted, what must operators do unless otherwise directed?
Repeat it at once from the manifold copy in the succession in which the several offices have been addressed, and then write the time of repetition on the order.

What should each operator receiving the order observe?
Whether the others repeat correctly.

After required signatures have been obtained to the order, to whom will the operator send these signatures, preceded by the order number?
To the train dispatcher, who will give "Complete," the time and the initials of whatever person is authorized to sign train orders.

What will each operator receiving this response then write on each copy?
The abbreviation "Com" for the word "Complete," the time and his last name in full.

To whom will he then deliver a copy?
To each person addressed, except that when the order is signed by the conductor, he must personally deliver a copy of it to the engineman.

When order is delivered by the conductor, what is required of the engineman before proceeding?
Some roads require that he read the order to the conductor.

When a "19" train order has been transmitted, what must operators do, unless otherwise directed?
Repeat it at once from the manifold copy, in the succession in which the several offices have been addressed.

What should each operator receiving the order observe?
Whether the others repeat correctly.

When the order has been repeated correctly, by an operator, what response will be given by the train dispatcher?
The word "Complete" and the time, with prescribed initials.

What will the operator receiving this response, then write on each copy?
The abbreviation "Com" for the word "Complete," the time, and his last name in full.

To whom will he personally deliver a copy without taking his signature?

To each person addressed.

When delivery to engineman will take the operator from the immediate vicinity of his office, by whom will engineman's copy be delivered?

By the conductor or brakeman.

When must a train be brought to a stop before delivery of a "19" train order?

When issued for it at the point where its superiority is restricted.

When may the train dispatcher OK the clearance under such circumstances?

After he has received definite advice from the operator that the train has stopped.

When so directed by the train dispatcher, how may a train order be acknowledged before repeating?

By the operator responding: *"Number of Train Order to Train Number X,"* with the operator's initials and office signal.

What must the operator then write on the order?

His initials and the time.

When may "Complete" be given to a train order for delivery to an inferior train?

Not until the order has been repeated or the X-response sent by the operator who receives the order for the superior train.

When a train order has been repeated, including the last word of the order, or X-response sent, and before "Complete" has been given, how must the order be treated?

As a holding order for the train addressed.

May it be otherwise acted on until "Complete" has been given?

No.

If the means of communication fails before an office has repeated an order or has sent the X-response, of what effect is the order at that office, and how must it be treated?

The order at that office is of no effect and must be there treated as if it had not been sent.

When clearing a train, how must the operator fill out clearance?

With proper address of train, and the numbers of all train orders addressed to the train, if any.

What must then be transmitted to the train dispatcher?

The address and order numbers from the clearance.

What will train dispatcher then do?

He will check, and if correct, will reply "OK," with the time and prescribed initials.

Will the operator then enter on the clearance the time and chief train dispatcher's initials.

Yes.

What record will the train dispatcher then make?

The record of orders with which train is cleared, and the time of OK will be recorded on clearance record.

When may train orders that have been repeated by the operator and made "Complete" by the train dispatcher, be delivered without the train dispatcher's OK to the clearance?

When the means of communication fails.

How will operator endorse such clearance?

"Wire failure."

When communication is restored, what will operator do?

Notify the train dispatcher of each train and time cleared. and the numbers of the train orders delivered.

Must operators make the requisite number of copies of clearance at one writing and preserve a copy?

Yes.

When clearances are received, and after checking them, what must conductors and enginemen know?

That their train is correctly designated, that orders are properly addressed, and that the order numbers on the clearance correspond with the orders received.

When additional orders are received after clearance has been repeated and OK response given, and before orders and clearance have been delivered, what must be done with first clearance?

It must be destroyed and a new one issued.

After clearances and orders have been delivered to a train and it becomes necessary to issue a restricting order to such train what must be done?

A Form 31 order must be used, the first clearance destroyed and a new one issued showing all orders delivered at that station.

If further orders, other than restricting orders, are issued after train has been cleared and it is not practicable to take up the first clearance, what must second clearance show?

All orders delivered and endorsed "Second" on top margin of clearance.

For train orders delivered by the train dispatcher, are the requirements as to the record and delivery the same as at other offices?

Yes.

When the train dispatcher issues a train order that is to be delivered to a train by the dispatcher, how must it be handled by him?

A carbon copy of the body of the order must be made in the dispatcher's train order book at the time the order is written on Form 19 or 31 and this copy underscored as required by Rule 206.

How must a train order, to be delivered to a train at a point not a train-order office, or at one at which the office is closed, be addressed?

To "C & E............at............care of............."

By whom will the order be forwarded and delivered?

By the conductor or other person in whose care it is addressed.

When Form 31 is used, upon whose signature will "Complete" be given?

The person by whom the order is to be delivered.

For whom must the person who is to deliver the order be supplied with copies?

The conductor and engineman addressed, an extra copy for the engineman, a copy for the rear trainman, and a copy upon which signatures of conductor and engineman shall be taken.

To whom must he deliver this copy?

To the first operator accessible.

What must the operator do with this copy?

At once transmit the signatures of the conductor and engineman to the train dispatcher, and preserve the copy.

Must orders delivered in this manner be acted on the same as if "Complete" had been given in the usual way?

Yes.

When a train passes a train-order signal at "stop" indication, what must it have before leaving the station?

A clearance.

When an operator at an intermediate station receives the signal "31" or "19" followed by the direction, what must he do?

Immediately display "stop" indication of train-order signal for the direction indicated and then reply "stop displayed," adding the direction.

When may the signal be restored to "proceed" indication?

Not until the orders have been delivered or annulled.

What signal appliances must operators have ready for immediate use?

One red flag, one white flag, one red lantern, one white lantern, six torpedoes, and three fusees.

Must operators promptly record and report to the train dispatcher the time of arrival and departure of all trains and the direction of extra trains?

Yes.

Must they observe trains and report at once to the train dispatcher if the proper signals are not displayed?

Yes.

When a train is named in an order by its schedule number alone, are all sections included?

Yes.

If an order is addressed to No. 5 and the dispatcher later runs three sections on this schedule, must each section be delivered a copy of the order?

Yes.

How long do train orders once in effect continue in effect?

Until fulfilled, superseded or annulled.

With what exception?

Orders held by or issued for, or any part of an order relating to a regular train, become void when such train loses both right and schedule as provided by Rules 4 and 82, or its schedule is annulled.

When is an order fulfilled?

When all its requirements have been completely carried out.

How may an order or part of an order be superseded?

By use of the words "instead of." (See Form P.)

How may an order be annulled?

By use of Train Order, Form L.

How may part of a train order be annulled?

By use of Train Order, Form M.

May any part of an order specifying a particular movement be either superseded or annulled?

Yes.

When a conductor or engineer, or both, are relieved before train orders in effect held by them have been fulfilled, what must be done.

Such train orders, clearances and instructions must be delivered to the relieving conductor and engineer.

Should such orders, clearances and instructions be read, checked and understood before proceeding?

Yes.

At registering stations what is required of operator in connection with train register?

If a train is to be registered by operator he must enter it first on the train register and then report it to the train dispatcher from the register. He must report the signals as registered by regular trains, "No signals," or "Green signals," as the case may be.

Should there be an error in registering a train or signals, should dispatcher require that it be corrected at once?

Yes.

At the time of relief, what is required of operators in making a transfer of the business?

The operator being relieved must call attention to overdue trains, trains which have been cleared, but are still at the station, and transfer all outstanding train orders and undelivered messages to the relieving operator on form provided, each operator signing the transfer.

Before signing the transfer what is required of the relieving operator?

He must check over all train orders carefully, noting the addresses in particular, and also inspect messages.

Must transfer be made in case an operator is relieved by another at meal time, or at any other time he may find it necessary to absent himself from duty for a time?

Yes.

At time of relief, may relieving operator handle train orders or the order-signal, as a matter of accommodation, until he has actually taken charge of the office and signed the transfer?

No.

How must transfer be made at offices which are not kept open continuously?

The operator going off duty must fill out the transfer, which must be signed by the operator who comes on duty later.

May an operator close an office, changing order signals to proceed, if he holds an order addressed to a train which is not due to arrive until the relieving operator is required to come on duty later?

The office must not be closed while any orders are in effect.

Why is it unsafe for two or more operators to handle train orders at an office during the same hours?

One operator may have cleared a train, and, not being aware of the fact, another operator might acknowledge an order restricting it; or one operator may give information to the dispatcher concerning the position of trains, which might prove misleading should another operator undertake any unfinished business.

Must operators keep dispatchers advised fully as to weather conditions and work necessary for trains to do at his station?

Yes.

At other than continuous offices, must operators keep a notice posted advising where they can be found during the hours the office is closed?

Yes.

What abbreviations may be used for:

Signature of train dispatcher?
His initials.

Conductor and engineman?
C&E.

Train will be held until train order is made "complete?"
X.

"Complete?"
Com.

Train report?
OS.

Number?
No.

Engine?
Eng.

Section?
Sec.

Passenger?
Psgr.

Freight?
Frt.

Minutes?
Min.

Junction?
Jct.

Train dispatcher?
Dispr.

Operator?
Opr.

Clear the line for train orders, and for operators to ask for train orders?
31 or 19.

"Stop displayed"?
S. D.

Division?
Div.

Subdivision?
Subdiv.

Miles per hour?
MPH.
Mile post?
MP.

May such office and other signals be used as are arranged by the superintendent of telegraph?

Yes.

May the usual abbreviations be used for the names of months?

Yes, except May, June and July.

May the words "Extra," "North," "East," "South," or "West," be abbreviated in train orders?

No.

TRAIN ORDERS

ORDER No. 1

No 16 Eng 25 meet No 15 Eng 30 at D

ORDER No. 2

Order No 1 should read No 16 Eng 20
instead of Eng 30

Under Order No. 1, if D is a blind siding and the engine on No. 2 is changed, what is required of the conductor?

He must explain to No. 15 the reason for No. 2 having an engine other than that mentioned in the order.

Must No. 15 wait until the correct engine number can be obtained by train order?

No. Having identified No. 16, that is all that is required.

If D is an open office, how should the error be corrected?

The dispatcher should correct as shown in Order No. 2.

ORDER No. 3

No 2 meet No 1 at C

ORDER No. 4

No 1 wait at D until 401 PM for No 2

If No. 2 arrives at C at 3:45 p.m. and can make B by 4:01 p.m., may it go there for No. 1?

No. Order No. 3 is in effect until fulfilled, superseded or annulled.

A——B——C——D——E——F——G——H
EASTWARD——→ ←——WESTWARD
EASTWARD TRAINS SUPERIOR BY DIRECTION

How should No 1 be governed under Orders Nos. 3 and 4?

They should wait at D until 4:01 p.m., then proceed to C and meet No. 2.

Should No. 2 start for D, knowing it could clear the 4:01 p.m. wait, and should No. 1 leave D ahead of the 4:01 p.m. wait, figuring on the meet to hold No. 2 at C, and a collision result, who would be responsible?

The crews of both trains.

Is a "positive meet" at C sufficient to hold No. 2 and render Order No. 4 invalid, inasmuch as No. 2 cannot possibly make D on the wait?

No. One train order is no more "positive" than another. No. 1 has a "positive wait" order which must be either fulfilled, superseded or annulled.

ORDER No. 5

No 2 wait at C until 3:01 PM for No 1

ORDER No. 6

No 2 meet No 1 at D

If No. 2 arrives at C at 2:40 p.m. and receives Order No. 6, may it proceed at once to D?

No.

Holding both orders, how should No. 1 be governed?

Remain at D and meet No. 2.

If a meeting point is fixed at the end of double track, in what way is the train affected which is going onto double track?

Should the train going onto double track arrive first, there is nothing to prevent it from proceeding onto double track without waiting, provided no order has been received singling the double track. The "right" form should be used in helping trains to the end of double track against trains going from double to single track.

Note: It will be remembered that form S-A applies only to

single track. Once the train moves from single to double track, its status is thereby changed to that of a double-track train and it is not then affected by single-track rules. However, should it again return to single track (as in the case of a short stretch of double track, say from C to D), its status then reverts to that of a single-track train and it is held by the meet order, provided it has not identified the train to be met while on double track.

Should double track extend only from C to D, may Extra 20 West and Extra 30 East be given an order reading: "Extra 20 West meet Extra 30 East on double track between C and D"?

No. Form S-A was not intended to be so used. The order should read: "Extra 20 West has right over Extra 30 East A to C not leave D unless Extra 30 East has arrived."

May the dispatcher fix a meeting point at the end of double track and in another order authorize the inferior train to run against the current of traffic, depending upon the meeting order alone to hold the ruling train at the designated place of meeting?

No. Having reached the end of double track, there is nothing to prevent a train from moving onto double track, moving with the current of traffic.

ORDER No. 7

No. 2 meet First and Second 1 at C

ORDER No. 8

No. 2 meet First 1 at D instead of C

After receiving Order No. 8, may No. 2 go to D for First 1, since Second 1 cannot pass First 1 at D in order to go to C on Order No. 7?

No.

When overtaken at D by Second 1, may First 1 arrange with it and proceed to C under protection of the orders held by Second 1?

No. No. 2 is tied up at C and First and Second 1 at D.

ORDER No. 9

Order No. 7 is annulled

A——B——C——D——E——F——G——H

EASTWARD——→ ←——WESTWARD

EASTWARD TRAINS SUPERIOR BY DIRECTION

How should No. 2 be governed after receiving Order No. 9?

Proceed to D and meet First 1, expecting Second 1 to clear No. 2's schedule time.

ORDER No. 10

No. 2 meet No. 1 at D and No. 3 at F

ORDER No. 11

No. 2 meet No. 1 at E instead of D

ORDER No. 12

Order No. 11 is annulled

How should No. 2 be governed under Orders Nos. 10, 11 and 12?

Run expecting No. 1 to clear its schedule time; proceed to F and meet No. 3.

ORDER No. 13

No. 2 meet No. 1 at C

ORDER No. 14

No. 2 meet First 1 at D instead of C

Under Order No. 13, how many sections of No. 1 must meet No. 2 at C?

All sections.

How should No. 2 be governed after receiving Order No. 14?

Remain at C. Order No. 14 establishes the fact that there are at least two sections of No. 1; and Order No. 14 does not supersede the meet with any except the first section.

If Order No. 2 had read, "No. 2 meet First and Second 1 at D instead of C," would Order No. 13 have been superseded?

Yes. No. 2 would not be expected to have knowledge of a third section of No. 1.

ORDER No. 15

No. 2 meet No. 1 at D

ORDER No. 16

No. 2 meet First 1 at C instead of D

Assuming there are four sections of No. 1, how should No. 2 be governed?

Meet First 1 at C and proceed to D and meet the three following sections.

If No. 2 were given an order to meet First, Second, Third and Fourth 1 at C and later be ordered to meet No. 1 at D instead of C, how many sections of No. 1 must No. 2 meet at D?

All sections.

ORDER No. 17

No. 2 meet No. 1 at C

ORDER No. 18

No. 2 meet No. 1 at E instead of D

Under Orders Nos. 17 and 18, may No. 2 leave C before the arrival of No. 1?

No.

When No. 1 receives Order No. 18, must it wait at E for No. 2?

Yes. Where the names of stations are similar, the crew on No. 2 might fail to catch the fact that Order No. 17 had not been properly superseded and move to E.

ORDER No. 19

No. 2 meet No. 1 at D and No. 3 at E
No. 2 take siding

ORDER No. 20

No. 2 meet No. 3 at F instead of E

Under Order No. 19, where must No. 2 take siding?

At both D and E. A better form of order would be: "No. 2 take siding and meet No. 1 at D and No. 3 at E."

Under Order No. 20, which train must take siding?

No. 3.

A——B——C——D——E——F——G——H

EASTWARD——→ ←——WESTWARD

EASTWARD TRAINS SUPERIOR BY DIRECTION

ORDER No. 21

No. 2 take siding and meet No. 1 at D
No. 3 at E and No. 5 at G

ORDER No. 22

No. 1 hold main track and meet No. 2 at
C instead of D

ORDER No. 23

No. 1 meet No. 2 at E instead of C

Under Order No. 21, where must No. 2 take siding?

At D, E and G.

Which train should take siding at C on Order No. 22?

No. 2.

Holding Orders Nos. 21, 22 and 23, how should No. 2 be governed at E?

Order No. 21 instructs No. 2 to take siding at E for No. 3; and Order No. 23 instructs No. 2 to hold main track at E for No. 1. This is an improper set of orders; and whatever moves No. 2 makes at E should be done under protection of flag. The dispatcher has No. 1 and No. 2 headed up in the siding.

ORDER No. 24

Eng 20 run extra A to D and meet Extra
30 West at E

Is Order No. 24 proper, even though the dispatcher intends to give Engine 20 a new running order from D to H?

No. Meeting points must be confined to the territory over which a train is authorized to run. On arrival at D, Order No. 24 becomes void to Extra 20 East, when that train ceases to exist; and if recreated, they cannot again use any part of Order No. 24.

If Extra 30 West held Order No. 24 and a running order H to A, would it be proper for the dispatcher to recreate Extra 20

East between D and H without reissuing the meet at E, depending on Order No. 24 to protect the movement?

Should the dispatcher issue a running order for Eng 20 between D and H without making provisions for that train to meet Extra 30 West, a lap of authority would exist. Should he issue an order: "Eng 20 run extra D to H and meet Extra 30 West at E," the movement would be protected after a fashion; however, this is a violation of Rule 202. The trains would be acting on orders under different numbers.

ORDER No. 25

Three Extras 20 5 and 210 East take
siding and meet Extra 40 West at D
take siding meet Extra 50 West at E and
hold main track and meet Extra 60 West at F

Why is it necessary to use the term "Three Extras"?

On roads where engine numbers run all the way from 1 to 1234, the train order could be so carelessly written that a misunderstanding would arise. For instance, if the order read: "Extras 20 5 and 210 East . . ." someone might misread it as Extras 205 and 210 East; or, with abnormal spacing Extra 1234 East might be misread as 12 and 34.

Is Order No. 25 perfectly clear?

No. Jostling along a rough-riding engine, trying to read this order quickly in a poor light, there is danger of misunderstanding. The majority of roads forbid the use of an order of this kind, where "take siding" and "hold main track" are mixed up in the same order. The same thing could be accomplished by a clearer wording, thus:

"Three Extras 20 5 and 210 East meet
Extra 40 West at D Extra 50 West at E
and Extra 60 West at F Three Extras 20 5
and 210 East take siding at D and E"

ORDER No. 26

Extra 20 West and No. 15 hold main
track and meet Extra 30 East at C
Extra 20 West run ahead of No. 15 H to A

Why should Order No. 26 instruct No. 15 to hold main track

A——B——C——D——E——F——G——H

EASTWARD——→ ←——WESTWARD

EASTWARD TRAINS SUPERIOR BY DIRECTION

when it is the superior train and would normally hold main track on a meet?

The order is so worded as to keep both westward trains on the main track for the eastward extra. A straight meet would have put Extra 30 East in the siding for No. 15 and Extra 20 West in the siding for Extra 30 East. A better form of order would have been:

> "No. 15 and Extra 20 West meet Extra 30 East at C Extra 20 West hold main track at C and run ahead of No. 15 H to A"

When a superior train is instructed to hold main track for an inferior train at a meeting point, there is always a question in the minds of the crew as to whether they may have missed an order somewhere which conferred right on the inferior train over them.

ORDER No. 27

> No. 16 Eng 40 meet First 15 Eng 20 at D and First 15 Eng 30 at E
> First 15 arriving E will be Second 15 leaving E

Under what circumstances would an order like No. 27 be necessary?

When a first section of No. 15 was to be started from E before the arrival of First 15 at E; making it necessary for No. 16 to meet two trains, both running as First 15.

ORDER No. 28

> No. 2 meet No. 5 at C and Second 5 at F
> There are two sections of No. 5 H to D

When would there be occasion for an order such as No. 28?

When First 5 takes down signals at D. Incidentally, this would not be permitted on most roads unless D was a register station. Signals are usually displayed through and Second 5 annulled D to A by using Form K.

ORDER No. 29

Extra 25 East with 75 cars back in and
meet No. 15 at C

How should both trains be governed under Order No. 29?

Extra 25 East may run by and back in at C, first protecting the movement as prescribed by Rule 99. No. 15 should allow ample room for the movement and must not come to the station until it is definitely known that Extra 25 East is in the siding. The extra should leave a flagman to inform No. 15 that the move has been made.

Should a later order be issued reading: "Extra 25 East meet No. 15 at B instead of C," what must be understood?

That Extra 25 East must pull into the siding at B.

ORDER No. 30

No. 2 meet No. 1 at B

ORDER No. 31

No. 1 meet No. 2 at C instead of B

ORDER No. 32

No. 2 meet No. 1 at B instead of C

Are Orders 30, 31 and 32 according to form?

Yes. However, such a set of orders is confusing. It would have been better for the dispatcher to annul Order No. 31 and then issue a meet at B.

ORDER No. 33

Extra 25 West meet Extra 30 East at B

ORDER No. 34

Extra 25 West meet Extra 30 East at C in-
stead of B

ORDER No. 35

Order No. 34 is annulled

After Order No. 35 is issued, where is the meet between the extras?

There is none.

A——B——C——D——E——F——G——H
EASTWARD——> <——WESTWARD
EASTWARD TRAINS SUPERIOR BY DIRECTION

If the extras collide because a lap of authority has been established by Orders 33, 34 and 35, who is responsible?

The train dispatcher.

ORDER No. 36

No. 16 meet No. 15 at C

ORDER No. 37

No. 15 has right over No. 16 C to A

After receiving Order No. 37, how should No. 15 be governed?

Remain at C and meet No. 16, as Order No. 37 does not supersede Order No. 36.

If the dispatcher wishes No. 15 to move C to A against No. 16 on Order No. 37, what must he first do?

Annul Order No. 36.

If No. 15 received Order No. 37 at D, together with an annulment of Order No. 36, how should they be governed?

If No. 16 is overdue, they should get out a flag immediately to protect against No. 16, as they have nothing on No. 16 between D and C. If they have time to move against No. 16's time-table time D to C, they could then use the right of track order to move against No. 16 from C to A.

What is the proper method of annulling the meet and moving No. 15 to A against No. 16 by an order issued at D?

The order should read: "Order No. 36 is annulled. No. 15 has right over No. 16 F to A," assuming that "F" is the last open office where an OS has been received on No. 15.

Does Order No. 36 confer right on No. 1 to move against No. 2 from where it receives the order to C?

Yes. That is the reason why the Form C order should confer right from the point where No. 1 was last reported.

ORDER No. 38

Order No. 36 is annulled
No. 15 has right over No. 16 D to A

Assuming that Order No. 38 is issued to No. 15 at D, if No. 15 were delayed between F (where last reported), and D, and No. 16 flags from A to D against No. 15, would there be anything to prevent No. 16 from leaving D and moving against No. 15?

No.

Should a collision result, who is responsible?

The dispatcher.

ORDER No. 39

No. 1 has right over No. 2 H to E

ORDER No. 40

No. 1 meet No. 2 at F

ORDER No. 41

Order No. 40 is annulled

What effect does Order No. 30 have on Nos. 1 and 2?

No. 1, inferior to No. 2 by direction, is made superior to No. 2 by right between H and E.

At what points does this order entitle No. 1 to hold main track?

At G and F; but not at E.

If the trains meet at E on Order No. 39, which train will hold the main track?

No. 2.

ORDER No. 42

Eng 25 run extra A to C and meet No. 15 at C

ORDER No. 43

Eng 25 run extra C to H and meet No. 15 at D

When Extra 25 East arrives at C, what happens to Order No. 42?

It is fulfilled and ceases to be effective. All orders held by the extra become void on arrival at C, excepting slow orders or annulments.

A——B——C——D——E——F——G——H

EASTWARD——→ ←——WESTWARD

EASTWARD TRAINS SUPERIOR BY DIRECTION

When they are recreated by Order No. 43, is there anything to prevent Extra 25 East from going to D to meet No. 15?

No.

Is Order 42 the best method of moving a train to the end of its run against a superior train?

No. Form C is preferable.

Holding both Orders 42 and 43, how should No. 15 be governed?

Remain at D and meet Extra 25 East.

ORDER No. 44

No. 1 has right over No. 2 F to C

If No. 2 makes some point between C and F and clears No. 1's schedule, what must they do?

Stop No. 1 and notify it of their arrival.

If No. 2 does not identify itself to No. 1, say at D, what effect does this have on No. 1?

On arrival at C, No. 1 must take siding, if they have not identified No. 2 between F and C; thus being delayed for a train they have already met.

ORDER No. 45

No. 1 meet No. 2 at D

ORDER No. 46

Order No. 45 is annulled

Holding Orders 44 and 45, how should Nos. 1 and 2 be governed?

They should proceed to C and there meet, as prescribed by the rules, No. 2 taking the siding.

When Order No. 46 is received, what effect does this have?

No. 2 loses its right to proceed to D against No. 1; and must then clear No. 1's schedule time, the same as before Order No. 45 was issued.

Does Order No. 45 supersede Order No. 44?

No.

ORDER No. 47

No. 15 has right over No. 2 H to C

ORDER No. 48

No. 15 meet No. 2 at D instead of having
right over No. 2 H to C

Is Order No. 48 a proper order?

No. When superseding an order, the dispatcher must stick to the same form as the order he is superseding. The meet at D could have been accomplished by an order reading: "No. 15 has right over No. 2 H to D instead of C;" or, by "No. 15 take siding and meet No. 2 at D"; or, by annulling Order No. 47, then fixing the meeting point at D, which would put No. 15 in the siding.

ORDER No. 49

First 15 has right over First 2 H to C

If First 2 makes some point between H and C for First 15, what must it do?

Stop First 15 and inform them of the arrival of First 2. Should they meet First 2 without identifying them, they would collide with Second 2.

When stopped by First 2 and notified of their arrival at D, in what position is First 15 placed?

First 15 is then on the main track at D on the time of an overdue superior train, Second 2.

If a dispatcher confers right on a train over one section of a schedule, must he confer equal, or more right on that train over all following sections?

Yes.

ORDER No. 50

No. 15 has right over First 16 H to F
and right over Second 16 F to C

ORDER No. 51

No. 15 meet First 16 at G

Holding both orders, after meeting First 16 at G, how should No. 15 be governed?

Remain at G for Second 16, since No. 15 has nothing against Second 16 between G and F.

A——B——C——D——E——F——G——H

EASTWARD⟶ ⟵WESTWARD

EASTWARD TRAINS SUPERIOR BY DIRECTION

If Second 16 could clear No. 15's time at F, would there be anything to prevent its leaving F and moving against No. 15?

No.

Should Second 16 strike No. 15 on the main track at G, who would be responsible?

The crew of No. 15 for not having protected their train, as prescribed by Rule 99. However, the dispatcher is primarily responsible for issuing orders in this manner. Order No. 50 should have read, ". . . and right over Second 16 H to C."

ORDER No. 52

No. 1 has right over No. 2 H to A

ORDER No. 53

No. 2 has right over No. 1 A to H

Is it proper for the dispatcher to attempt to nullify the provisions of Order No. 52 by an order such as No. 53?

No. Order No. 52 should have been annulled, thus re-establishing No. 2's superiority by direction.

ORDER No. 54

No. 2 meet First 1 at D

ORDER No. 55

Second 1 has right over No. 2 H to C

If No. 2 receives Order No. 55 at C, how should it be governed?

Proceed to D, protecting against Second 1 between switches at D, as per Rule 99. After meeting First 1 at D, they would then be on the main track on the time of an overdue superior train.

Note: On roads where the Second paragraph of Rule 94 is still retained, it is usually required that No. 2 not leave C, as there is the danger of Second 1 overtaking First 1 at D and taking them ahead of them from D to C under the provisions of this rule.

ORDER No. 56

Extra 30 East has right over No. 1 A to D

ORDER No. 57

Extra 30 East wait at C until 401 PM for
No. 1

Under Order No. 56, may No. 1 leave D before Extra 30 East arrives?

No.

If No. 1 can make C on the 4:01 p.m. wait and clear the extra as much as the extra was before required to clear No. 1, may they go?

Yes.

Must No. 1 stop Extra 30 East at C and notify it of their arrival?

Yes, to prevent the extra from being delayed at D for a train it had already met.

May an order be issued reading, "No. 1 has right over No. 2 to C"?

No. Both stations between which right is given should be stated.

May an order be issued reading: "Eng 20 run Extra H to A with right over No. 16"?

Some roads permit the use of this form. However, it is not authorized by the Standard Code.

ORDER No. 58

No. 15 has right over No. 16 H to D and
at D

How should No. 16 be governed on arrival at D?

Unless No. 16 can make some point beyond D and clear No. 15's time, they should take siding at D, as this order entitles No. 15 to hold main track at D.

Note: This is an order seldom found authorized in a rule book.

A——B——C——D——E——F——G——H
EASTWARD——→ ←——WESTWARD
EASTWARD TRAINS SUPERIOR BY DIRECTION

ORDER No. 59

Eng 30 run extra A to H with right over
No. 15 and meet Extra 20 West at D

Is Order No. 59 in proper form?

Most roads permit the use of a combination of forms, such as this. Others object on the grounds that if the dispatcher wishes to annul some movement contained in the order, he may inadvertently annul the entire order, thus leaving the extra on the main track without authority to move, its running order having been annulled.

ORDER No. 60

No. 2 run 30 minutes late C to H

How late must No. 2 pass C, D, E, F and G?

Not less than thirty minutes late.

If No. 2 is due at F at 3:15 p.m. on the time-table, when must No. 1 be clear at F under Order No. 60?

They must be clear of the main track before No. 2's leaving time, 3:45 p.m.

What time would a second-class train be required to clear?

A second-class train would clear the same leaving time; however, they must be in the clear at least five minutes before No. 2 is due to leave, or at 3:40 p.m.

ORDER No. 61

No. 2 run 30 minutes late A to D and 10
minutes late D to H

How late must No. 2 pass each station?

Thirty minutes late at A, B and C; and ten minutes late at D, E, F and G.

May No. 2 arrive at D less than 30 minutes late?

Yes.

When could the dispatcher start a First section of No. 2 from D?

Ten minutes late on its schedule.

If the dispatcher wished to move an inferior train against No. 2 to D by giving them thirty minutes of No. 2's time at that point, how should it be done?

By the use of a wait order, thus: "No. 2 run 30 mins late A to D and wait at D until 430 PM and run ten mins late D to H," assuming No. 2's schedule time at D to be 4:00 p.m.

ORDER No. 62

No. 1 run fifty 50 mins late H to A

ORDER No. 63

No. 1 run thirty 30 mins late H to A

Under Orders 62 and 63, how late must No. 1 run?

Fifty minutes late.

If given both orders, how much of No. 1's time could Extra 22 East use?

Thirty minutes. Order No. 63 does not supersede Order No. 62; however, there is always the chance that the dispatcher has annulled Order No. 62 to No. 1 in error.

ORDER No. 64

No. 1 run one 1 hour late H to A

ORDER No. 65

No. 1 run two 2 hours late H to A

Given Orders Nos. 64 and 65, how much of No. 1's time could Extra 22 East use? Why?

Two hours. It is neither proper, nor necessary to supersede a time order; and where a later order increases the run-late, this is proper and customary. No. 1 must run two hours late. It is only when time is reduced, as in Orders 62 and 63 that there is doubt.

Is Order No. 64 fulfilled by No. 1 when they comply with Order No. 65?

Yes.

If Order No. 65 were annulled, how late would No. 1 run?

One hour late.

If an operator receives Orders Nos. 64 and 65, must both be delivered to No. 1?

Yes.

A——B——C——D——E——F——G——H

EASTWARD——> <——WESTWARD

EASTWARD TRAINS SUPERIOR BY DIRECTION

If Order No. 65 were addressed to "Yardmaster," or "Yard Engines" at a terminal, is it safe for the dispatcher to reduce the run-late time by annulling Order No. 65? Why?

No. This would amount to restricting yard movements; and the rules do not provide the dispatcher with such authority.

ORDER No. 66

Second 1 run thirty 30 mins late H to A

How late will First 1 run under Order No. 66?

On time.

Should a dispatcher use the run-late order to move inferior trains against a schedule train if it can be avoided?

Not if the train can make up time on its schedule. A "wait order" is preferable; and some roads do not permit the use of the run-late at all.

ORDER No. 67

First and Second 1 run thirty 30 mins late
H to A

If an inferior train makes F under Order No. 67, and No. 1 arrives with NO signals, may the train proceed?

It would not be safe for the inferior train to proceed without definitely knowing that the train met was fulfilling the entire schedule. The rules of most roads prohibit the dispatcher from taking down signals at an intermediate point, requiring that signals be displayed and the second section then be annulled by the use of Form K.

ORDER No. 68

No. 2 wait at C until 1001 AM for No. 1
and Extra 20 West

If No. 1 goes to C for No. 2, what time must they be in clear?

Not later than 10:01 a.m. The rules of most roads require a one-minute clearance, or 10:00 a.m. to take care of the allowable variation in watches.

When must Extra 20 West be into clear?

Not later than 9:56 a.m.

If unable to make C for No. 2, can Extra 20 West use the wait to go to D?

Yes.

If No. 1 arrives at C and finds it has time to go to B and clear No. 2's time-table time, may it go?

Yes.

Under what circumstances may No. 2 leave C before 10:01 a.m.?

Only if No. 1 and Extra 20 West have arrived.

If Extra 20 West goes to D for No. 1, being unable to make C, what time must it be into clear?

Not later than 9:56 a.m., the same time it would clear at C.

ORDER No. 69

No. 2 wait at C until 1130 AM

When may No. 2 leave C?

At 11:30 a.m.

May any inferior train receiving Order No. 69 use the 11:30 a.m. wait to go to C for No. 2, or any intermediate station between H and C where the 11:30 a.m. wait is later than the schedule time?

Yes.

When may the dispatcher start a first section of No. 2 from C, D, E or F?

Not before 11:30 a.m.

ORDER No. 70

First 2 wait at D until 1130 AM
Second 2 wait at C until 1150 AM for
No. 15

When may First 2 leave D?

Not before 11:30 a.m.

When may Second 2 leave C?

Not before 11:50 a.m., unless No. 15 has arrived.

A——B——C——D——E——F——G——H
EASTWARD——→ ←——WESTWARD
EASTWARD TRAINS SUPERIOR BY DIRECTION

If First 2 arrives at D at 11:20 a.m. and finds No. 15 there, may they proceed?

No. Any train receiving Order No. 70 could use the 11:30 a.m. wait to move against First 2 to D.

Should the dispatcher issue an order of this kind.

No. It is confusing.

ORDER No. 71

No. 2 wait at D until 1130 AM for No. 15

ORDER No. 72

No. 2 run one 1 hour late D to H

Assuming No. 2's schedule time at D to be 10:01 a.m., when may they leave D under Orders 71 and 72?

Not before 11:30 a.m., unless No. 15 has arrived. If No. 15 is found at D, then No. 2 may leave any time after 11:01 a.m., one hour late.

Holding both orders, how should No. 15 clear?

They must be into clear at D not later than 11:25 a.m.; or they may use the 11:30 a.m. wait, less the required five minute clearance, at any point between H and D where the wait is later than the one-hour run-late.

ORDER No. 73

No. 2 run one hour late D to H

ORDER No. 74

No. 2 wait at D until 1050 AM for No. 15

Again assuming No. 2's schedule time at D to be 10:01 a.m., when should No. 15 clear at D?

Not later than 10:45 a.m. Order No. 74 does not supersede Order No. 73, which permits No. 2 to leave D at 11:01 a.m. However, such a combination of orders creates doubt; and the safe course is to clear the earlier time.

If an operator receives a middle order addressed only to the

"Operator," reading: "No. 2 wait at D until 1050 AM," when may he change his order board to "proceed"?

After the order has been delivered to No. 2, regardless of time. After 10:50 a.m., the dispatcher may annul the middle order, but the operator cannot file it unless annulled.

ORDER No. 75

No. 2 take siding at C
No. 2 wait at C until 955 AM for No. 15

How should No. 15 be governed under Order No. 75?

No. 15 must arrive at C not later than 9:50 a.m. and, under the provisions of the order, will hold the main track.

If No. 2 arrives at C at 10:15 a.m., how may they proceed?

They may proceed directly through the siding at any time after 9:55 a.m. but may not proceed up the main track at any time except under protection of flag, even though No. 15 is not found at C.

Should the dispatcher limit the use of this order to cases where no other order will accomplish the same purpose?

Yes. The rules of most roads do not authorize this form. There are times when such an order will get a short-time crew into a terminal before expiration of their 16 hours, when the dispatcher would not be using good judgment to put No. 2 in the siding on a meet order. Given a meet, No. 2 would be indefinitely delayed at C should the crew of No. 15 tie up before arrival at C.

If No. 2 meets No. 15 at some siding before reaching C, must it still take siding at C?

Yes, or flag up the main track between switches.

ORDER No. 76

No. 2 wait	A until	805 AM
	B	825 AM
	C	901 AM
	D	930 AM
	E	1001 AM

ORDER No. 77

No. 2 run twenty 20 mins late on Order No. 76

When may No. 2 leave A, B, C, D and E?

A——B——C——D——E——F——G——H

EASTWARD⟶ ⟵WESTWARD

EASTWARD TRAINS SUPERIOR BY DIRECTION

Twenty minutes later than the times shown in Order No. 76: A 8:25 a.m., B 8:45 a.m., C 9:21 a.m., D 9:50 a.m. and E 10:21 a.m.

If No. 15 were at F, what time must it clear No. 2 at E?

At 10:16 a.m.

If No. 15 were at G, could it use the same time to clear No. 2 at F?

Yes.

SCHEDULE OF No. 2

A	805 AM
B	825 AM
C	901 AM
D	950 AM
E	1001 AM
F	1015 AM
G	1030 AM
H	1045 AM

ORDER No. 78

No. 2 run twenty minutes late A to E

What time must No. 15 clear No. 2 at E under Order No. 78?

Must be into clear not later than 9:56 a.m.

When may the dispatcher start a first section of No. 2 from E?

On time, 10:01 a.m.

When may No. 2 leave D?

Not less than 20 minutes late, 10:10 a.m.

If No. 15 is at F at 9:50 a.m. and unable to clear No. 2's schedule time at E, may they use the 10:10 a.m. time at D to move against No. 2 from F to E?

No.

ORDER No. 79

Eng 20 display signals and run as First 1
H to A

ORDER No. 80

Eng 25 run as Second 1 H to A

How should order No. 79 be addressed?

To Engine 20.

How should Order No. 80 be addressed?

To Engine 25.

Should Second 1 be given a copy of Order No. 79, authorizing the display of signals for them?

No.

ORDER No. 81

Second 1 display signals F to C for Eng 99

How should Order 81 be addressed?

To Second 1 and Engine 99.

ORDER No. 82

Engs 20 25 and 99 run as First Second
and Third 1 H to A

Under Order No. 82, which engines will display signals?

Engines 20 and 25.

ORDER No. 83

Eng 85 display signals and run as Second
1 F to A

Following sections change numbers accordingly

How should Engine 85, Second and Third 1 be governed at F?

Engine 85 should display signals and run as Second 1 F to A. Second 1 into F would be Third 1 leaving F, with green signals; Third 1 arriving F with no signals would leave as Fourth 1 with no signals.

ORDER No. 84

Eng 85 is withdrawn as Second 1 C to A

How would Second, Third and Fourth 1 be governed at C under Order No. 84?

A——B——C——D——E——F——G——H
EASTWARD⟶ ⟵WESTWARD
EASTWARD TRAINS SUPERIOR BY DIRECTION

Second 1 would drop out at C. Third 1 into C would take the next lowest number and be Second 1 leaving C with green signals. Fourth 1 with no signals would become Third 1 with no signals out of C.

In adding, or withdrawing sections of a schedule, must the dispatcher provide all orders essential to their movement beyond the point where the order of sections are changed?

Yes.

Is it proper to use an order such as No. 84 to withdraw a first section?

No. The original signal order should be annulled and a new one issued.

Is it proper to discontinue a last section by instructing the preceding section to take down signals at an intermediate point?

No. If the last section is to be discontinued, signals should be displayed through by the leading section and the last section annulled by the use of Form K.

ORDER No. 85

Eng 18 instead of Eng 85 display signals
and run as Second 1 D to A

When is an order such as No. 85 necessary?

When an engine handling a schedule is to be substituted for the engine then handling the train. Where more than one section has been created in a signal order at a terminal and it is not feasible to annul the original signal order due to an engine change, this form may be used.

If the engine change involves a last section, will the words "display signals and" be omitted from the form?

Yes.

Is it necessary to address such an order to other sections of the schedule?

No.

ORDER No. 86

Second 1 take down signals at D

Is it proper to discontinue a last section by the use of this form?

Some roads permit it. However, the majority of roads require that Second 1 continue the display of signals through to A; and that Third 1 be annulled D to A.

ORDER No. 87

Engs 99 and 25 reverse positions as
Second and Third 1 D to A

How should Second and Third 1 be governed at D?

Second 1 will allow Third 1 to pass, first exchanging orders, numbers and signals. That is, Third 1 with no signals becomes Second 1 with green signals; and Second 1 with green signals becomes Third 1 with no signals.

To whom should Order No. 87 be addressed?

To C&E Second and Third 1.

If there were more than three sections of No. 1, should following sections be given a copy of the order?

No.

If engine numbers are used in train orders, can confusion be eliminated by addressing a copy of Order No. 87 to those trains who hold orders against Second and Third 1?

Yes.

ORDER No. 88

Engs 40 20 and 30 display signals and
run as First Second and Third 2 A to H

Which sections will display signals under Order 88?

All three sections.

What is the use of such an order?

It may be used when four or more sections are to be run on a schedule and only the engines of the first three sections are known.

ORDER No. 89

Eng 66 run as Fourth 2 A to H

Is Order No. 89 all that Fourth 2 needs, or must they have

A——B——C——D——E——F——G——H
EASTWARD——→ ←——WESTWARD
EASTWARD TRAINS SUPERIOR BY DIRECTION

a copy of the order authorizing the first three sections?
Order No. 89 is all that is required.

ORDER No. 90

Order No. 88 is annulled
Engs 20 30 and 40 display signals and run
as First Second and Third 2 D to H

When the order of sections is altered by Order No. 90, do all orders held by the three trains become void?

Yes, with the exception of slow orders and annulments.

Must the dispatcher provide First, Second and Third 2 with new orders, with the exception of slow orders and annulments?

Yes, with this exception. Some roads provide that where the trains have orders in their possession which are made ineffective, as in this case, the dispatcher may reinstate their effectiveness by an order reading: "Respect Orders No. ...." Usually, a train may be instructed to respect only a helping order; and a restricting order must be reissued to the train by the dispatcher, even though the crew has a copy of it.

If the dispatcher runs sections of a train to an intermediate point of a schedule, must he specify which section is to fulfill the schedule from that point?

Yes.

How should this be done?

By an order such as the following: "Engs 21 and 22 run as First and Second 2 A to D Eng 22 run as No. 1 from D."

ORDER No. 91

Address: C&E No. 2 and Eng 30 at C
No. 2 display signals C to D for Eng 30

When would this form of order be necessary?

When No. 2 has been created by a clearance and it is desired to have them display signals later.

ORDER No. 92

Engs 20 and 25 display signals and run as
First and Second 2 A to H

ORDER No. 93

Engs 20 and 25 run as First and Second 2
A to H

Explain the difference in Orders 92 and 93?

Under Order No. 92, both sections of No. 2 would display signals. Order No. 93 does not authorize a third section; therefore, only First 2 would display signals.

How are trainmen and enginemen to know that the section ahead has gone and that it displayed signals for a following section?

By checking the train register, if there be one. Otherwise, it becomes the duty of the dispatcher to issue necessary instructions to prevent sections from running out of sequence, if the leading section cannot be plainly seen.

ORDER No. 94

Third 1 due to leave A Feb 22 has arrived at F with no signals and is annulled
F to A

Why is it necessary that Order No. 94 state "with no signals," or "with green signals," as the case may be?

Opposing inferior trains have no way of knowing how many sections of No. 1 were started from H; therefore, if the order did not state "with no signals," there could be a Fourth 1 authorized between H and F that had not arrived at F. Should the inferior train take it for granted that Third 1 was the last section and collide with Fourth 1 between F and H, the crew of the inferior train would be held responsible. If no register is maintained at F, the order must state whether the section annulled was displaying signals.

Should signals be taken down at an intermediate point before the section for which they were displayed arrives, what is required of the dispatcher?

Opposing inferior trains must be notified at or before reaching point where signals were taken down, if there be no register

at that point. The majority of roads do not permit taking down signals except at register stations.

Should the dispatcher instruct No. 1 to display signals F to A for Engine 99, must inferior trains following No. 1 be advised?

Yes, until such time as Second 1 has left F.

ORDER No. 95

Eng 20 run extra A to D
No. 1 run one 1 hour late H to A

ORDER No. 96

Eng 20 run extra D to H

When Extra 20 East arrives at D and is given Order No. 96, must they be given the time on No. 1 again, or may they use Order No. 95?

Order No. 95 expired on their arrival at D and they may use no part of it. The dispatcher must reissue the time on No. 1.

If Extra 20 East is the only train holding the one-hour run-late on No. 1, may the dispatcher annul the restriction to No. 1 after Extra 20 East has arrived at D?

Yes.

ORDER No. 97

Eng 20 run extra A to D and return to A
No 1 run 2 hours late H to A

Under Order No. 97, if Extra 20 East finds it impracticable or unnecessary to go to D, may it use the return portion of the order without having done so?

No.

Upon arrival at D, must Extra 20 East pull into the siding; and should a clearance card be obtained before starting the return trip, if it be an open train-order office?

Yes.

May the dispatcher place a restricting order at D, depending on the crew to obtain a clearance before starting the return movement?

No.

May the extra use the time on No. 1 in both directions?

Yes. It is a continuous movement and the time on No. 1 may be used as long as Order No. 97 is in effect.

Should it be desired to turn Extra 20 East back from B instead of permitting them to go to D, should Order 97 be annulled and a new running order issued?

Yes.

If Extra 20 East is given an order to meet a westward extra at D, may it return from D to A without waiting to meet the westward extra at D?

Yes. However, Form A is not a proper order for such a move. Form C is preferable.

ORDER No. 98

Eng 30 run extra A to D and return to A
Eng 20 run extra A to D with right over
Extra 30 West

May Extra 30 leave D on its return, or westward trip before the arrival of Extra 20 East?

No.

If Extra 20 East is ahead of Extra 30 East, could the movement be safeguarded by directing Extra 30 East not to pass Extra 20 East, instead of the manner handled in Order No. 98?

No.

ORDER No. 99

Eng 20 has until 950 AM to run extra
C to D with right over Westward Trains

ORDER No. 100

Eng 20 run extra C to D and has right
over Westward Trains C to D
This order is annulled at 950 AM

Are the meanings of Orders Nos. 99 and 100 identical?

Yes.

A——B——C——D——E——F——G——H

EASTWARD——→ ←——WESTWARD

EASTWARD TRAINS SUPERIOR BY DIRECTION

When may any westward train leave D?

After the arrival of Extra 20 East, or if the extra does not arrive, then after 9:50 a.m.

Failing to make D by 9:50 a.m., how should Extra 20 East be governed?

It should protect in both directions, as prescribed by Rule 99.

Is it understood that Extra 20 East loses all authority to occupy the main track at 9:50 a.m., and must be properly protected prior to the expiration of that time?

Yes.

ORDER No. 101

After 210 PM Eng 20 run extra A to H

When does Extra 20 East have authority to occupy the main track at A?

Not until after 2:10 p.m.

When is such an order necessary?

When some opposing, or conflicting movement, is limited by time, i.e., "Eng 21 has until 205 PM to run extra B to A."

ORDER No. 102

After Extra 21 West arrives at A Eng 30
run extra A to H

Is Order No. 102 valid to Engine 30 before the arrival of Extra 21 West?

No.

If given an order, "Extra 30 East meet Extra 21 West at B," would this validate the order and permit Extra 30 East to leave A?

No. That part of Order No. 102 reading "After Extra 21 West arrives at A" must first be annulled.

ORDER No. 103

Eng 20 run extra A to E and meet No. 15
at E

ORDER No. 104

Eng 20 run extra E to H and meet No.
15 at F

If, on arrival at E, Extra 20 East is given Order No. 104, may they go to F for No. 15?

Yes.

Should Order No. 104 read "instead of E"? Why?

Order No. 103 becomes invalid to Extra 20 East on their arrival at E and they may use no part of it. Therefore Order No. 104 should not read "instead of E," as Extra 20 East no longer holds a meet with No. 16 at E.

ORDER No. 105

Eng 21 run extra H to A
Eng 22 run extra A to D and meet Extra
21 West at F

ORDER No. 106

Eng 22 run extra D to H

When does Order No. 105 become void to Extra 22 East?

On their arrival at D.

Does that portion of the order providing for Extra 22 East to meet Extra 21 West at F remain in effect?

No. The order becomes void to Extra 22 East in its entirety.

After given Order No. 106, where do the two extras meet?

There is no meet, insofar as Extra 22 East is concerned.

Does Extra 21 West still have a meet with Extra 22 East at F?

Yes, since the meet lies within the limits of its running order.

Should Extra 22 East pass F before Extra 21 West arrives and the trains collide, who would be responsible?

The dispatcher.

Would it be good judgment for the crew on Extra 22 East to inquire about Extra 21 West when given their new running order?

Yes.

A——B——C——D——E——F——G——H
EASTWARD——→ ←——WESTWARD
EASTWARD TRAINS SUPERIOR BY DIRECTION

Is it improper to give an extra a meet with a train beyond the point to which it is authorized to move?

Yes.

ORDER No. 107

Engs 20 and 30 run as two extras A to D
and return to A

Is there anything in Order No. 107 to prevent either of the extras from returning from D before the other arrives?

No. The order constitutes a lap of authority.

How should the order read if both extras are to go to D before either begins the return trip?

"Extras 20 and 30 East have right over Extras 20 and 30 West A to D."

How long does an extra's running order continue in effect?

If not fulfilled, it continues in effect indefinitely.

If called to leave A on July 1 and given a running order dated June 30, may the extra use the order?

Yes.

When is an extra's running order fulfilled?

When the extra arrives at the station last named in the order.

ORDER No. 108

No. 2 run thirty 30 mins late A to H

ORDER No. 109

After helping No. 44 A to C Eng 21 run
extra C to A respecting Order No. 108

May Extra 21 West use the 30-minute run-late on No. 2 from C to A?

Yes, provided the rules of the road authorize an order such as No. 109. Order No. 108 is a helping order to Extra 21 West; and some roads permit the dispatcher to validate an order in this manner when the crew holds the order in their possession but, under the rules, cannot use it. It saves time for the dispatcher, since he does not have to reissue the order to the extra at C when they previously received it when helping No. 44.

ORDER No. 110

Eng 20 run extra H to A

ORDER No. 111

Eng 30 run extra A to H and wait at C
until 1001 AM for Extra 20 West

If Extra 20 West arrives at C at 9:50 a.m., is there anything to prevent its continuing on its running order?

No.

If Extra 20 West does not make C by 10:01 a.m., is there anything to prevent its continuing on its running order to A?

No.

Has Extra 30 East been made superior to Extra 20 West?

No.

If a collision results, who is responsible?

The dispatcher.

How should Order No. 111 have been worded to avoid this lap of authority?

"Eng 30 run extra A to H with right over Extra 20 West and wait at C until 1001 AM for Extra 20 West."

Is there any superiority between extra trains, except that provided for by train order?

No.

ORDER No. 112

Eng 20 run extra A to C

ORDER No. 113

Eng 30 run extra H to C

Do Orders Nos. 112 and 113 constitute a lap of authority?

No. Extra 20 East must head in at the west switch at C. Extra 30 West must head in at the East switch at C. The rules provide that each shall move on the siding in such a manner that a collision will not result.

ORDER No. 114

Eng 20 run extra A to H
Eng 30 run extra H to A and meet
Extra 20 East at D

A——B——C——D——E——F——G——H

EASTWARD——→ ←——WESTWARD

EASTWARD TRAINS SUPERIOR BY DIRECTION

Which extra will take siding at D?

The extra in the inferior time-table direction, Extra 30 West.

Does this imply superiority between the two extras?

No. The meet makes each superior to the other to D; and the rules provide that the extra in the inferior time-table direction shall take the siding unless the order states otherwise.

ORDER No. 115

Eng 30 run extra A to H and has right
over Extra 40 West A to D

On arrival at D, if Extra 40 West is not there, is there anything to hold Extra 30 East from continuing on its running order.

There are two answers to this question. Unless the rules of a road specifically state that Extra 30 East shall take siding and not leave D unless Extra 40 West has arrived, then the order should so state, as there would be nothing to prevent Extra 30 East continuing on its run. On roads having a rule putting Extra 30 East in the siding and holding it at D, the order is proper. On roads not having this rule, the order constitutes a lap of authority.

ORDER No. 116

Eng 20 run extra leaving A Thursday Jan
1st as follows with right over all trains

Leave	A	eleven thirty	1130 PM
	B	eleven forty five	1145 PM
	C	twelve one	1201 AM
	D	twelve twenty	1220 AM
Arrive	E	twelve thirty	1230 AM

Does Rule 82 apply to Order No. 116?

No.

How long does Order No. 116 continue in effect?

Until Extra 20 East arrives at E, or the order is annulled.

How many minutes must opposing train clear the time of Extra 20 East?

Not less than five minutes.

May an exception be made of certain regular trains, and if any are excepted, how must the extra clear those trains?

The order may be varied to except certain trains. If, for instance, the order conferred right over all trains except first-class, the extra would be required to clear first-class trains the same as any other extra.

May the dispatcher complete the order to Extra 20 East before it has been made effective to all trains over which right is given?

No.

Is such an extra relieved from compliance with Rule 99, if delayed?

No.

How should Extra 20 East move through yard limits?

As prescribed by Rule 93, the same as any other extra train.

May Extra 20 East be given right over "Yard Engines"?

No. The dispatcher does not create yard movements; and the rules make no provisions for his restricting them.

If engines 20, 30 and 40 were run as three extras on one order of this kind with right over all trains, would this be proper?

It would not only be improper but impossible. If the three extras each had right over each other, none of them could move; since, when one was due to leave A, all would be due to leave, and each extra would be required to clear the other two. The order could be made to read: ". . . with right over all trains except each other"; however, the occasion when such an order is required seldom arises.

Must trains in the same direction clear Extra 20 East in the same manner as required for a first-class train?

The rules of most roads so provide.

Must work extras clear the time of Extra 20 East, whether or not they have been instructed previously not to protect against extra trains?

Yes.

If an opposing extra were created with the same form of

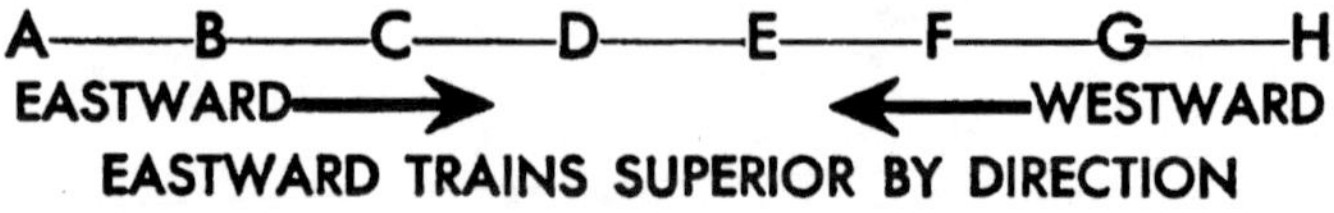

order, should Extra 20 East be excepted from the trains over which right is given?

Yes.

If Engines 20, 30 and 40 were run on a single order, and an opposing train received an order that Extra 20 East would wait at C until 12:30 a.m., may it go to C for Extras 30 and 40 East on the assumption that sequence must be maintained between the extras?

No.

May Extra 20 East be given an order to run late on Order No. 116?

The rules of most roads provide for this; however, there are roads which do not permit the use of any kind of a run-late order.

Must time be given at each station within the limits of such an order?

This is optional with the dispatcher.

ORDER No. 117

Eng 25 works extra 701 AM until 601
PM between C and D

How should Work Extra 25 be governed under Order No. 117?

It must clear the time of regular trains and protect itself against extra trains in both directions.

ORDER No. 118

Eng 25 works extra 701 AM until 601
PM between C and D not protecting
against Eastward Extra trains

How should Work Extra 25 be governed under Order No. 118?

The time of regular trains must be cleared and protection must be provided against westward extra trains.

ORDER No. 119

Eng 25 works extra 701 AM until 601 PM between C and D not protecting against extra trains

ORDER No. 120

Work Extra 25 protects against Extra 40 East between C and D after 130 PM

ORDER No. 121

Work Extra 25 protects against Second and Third Class Trains between C and D

Holding Orders Nos. 119, 120 and 121, how would Work Extra 25 be governed?

First-class trains must be cleared; the work extra could work upon the time of second- and third-class trains under protection of flag, instead of clearing them; and extra trains would not be protected against until 1:30 p.m., when the work extra must have out a flag against Extra 40 East. After Extra 40 East has gone, the work extra may then work without further flag protection against extra trains.

ORDER No. 122

Work Extra 25 clears Extra 50 West after 230 PM

May Work Extra 25 occupy the main track after 2:30 p.m., flagging against Extra 50 West?

No. It must clear the main track before 2:30 p.m. Only in emergency may it occupy the main track later than that time under flag protection.

Going back to Order No. 121, how will second- and third-class trains run between C and D?

Expecting to find Work Extra 25 working on the main track under protection of flag.

Under Order No. 120, may Extra 40 East enter the working limits prior to 1:30 p.m. under protection of flag?

No.

How may the limits of a work extra, or its time be extended?

By annulling the original work order and issuing a new one.

A——B——C——D——E——F——G——H
EASTWARD——→ ←——WESTWARD
EASTWARD TRAINS SUPERIOR BY DIRECTION

ORDER No. 123
No. 62 meet Work Extra 25 at C

If No. 62 holds Orders Nos. 119 and 123 and is at C at 6:01 p.m. and Work Extra 25 has not arrived, may No. 62 go?

No.

If Work Extra 25 is given a new working order after 6:01 p.m., could it use Order No. 123 to go to C for No. 62?

No. All of its orders, with the exception of slow orders and annulments, become void when its work order expires at 6:01 p.m.

Under Order No. 119, may the work extra occupy the main track between switches at either C or D?

No.

ORDER No. 124
Eng 40 works extra 701 AM until 601 PM between C and E clears Extra 21 West protecting against second and third class trains not protecting against extra trains except protects against Extra 30 East after 1030 AM and Extra 35 West after 1230 PM

Does Order No. 124 conform to the Standard Code?

No, although the rules of many roads permit the use of such an order.

How should Work Extra 40 be governed?

It must not enter the main track until after Extra 21 West has been met. It may work on the time of second- and third-class trains, clearing the time of other regular trains; but must protect itself against overdue second- and third-class trains. Protection against eastward extra trains is not required until 10:30 a.m. when it must have out a flag against Extra 30 East. It will work without flag protection against westward extra trains until 12:30 p.m., when it must have out a flag against

Extra 35 West. Having met Extra 30 East and Extra 35 West, no further flag protection is required against extra trains.

If later given an order reading: "Work Extra 40 protects against Extra 35 West between C and E after 1130 AM," must Work Extra 40 have out a flag against the extra by 11:30 a.m.?

Yes.

If given a copy of the same order, when may Extra 35 West enter the limits of the work extra?

Not before 12:30 p.m. The order should have read ". . . after 1130 AM instead of 1230 PM."

ORDER No. 125

Engs 20 and 21 work as two work extras
801 AM until 201 PM between C and E
not protecting against extra trains

Does Order No. 125 require protection between Work Extras 20 and 21?

No.

Should a collision result, who is responsible?

The dispatcher.

How should the order have read?

It should have provided for one of the work extras to protect against the other, i.e., ". . . not protecting against extra trains except Work Extra 20 protects against Work Extra 21."

Would the same result be accomplished by making the order read ". . . except protect against each other"?

No. If one work extra protects against the other, this leaves the other free to move. Where two work trains protect against each other, neither is free to move against the other except under protection of flag.

If, after clearing an extra it has been instructed to clear, the work extra overtakes the extra cleared, do the provisions of the order again become effective?

Yes. The work extra must get in the clear and stay clear of the extra throughout the entire working limits.

A——B——C——D——E——F——G——H

EASTWARD⟶ ⟵WESTWARD

EASTWARD TRAINS SUPERIOR BY DIRECTION

ORDER No. 126

Work Extra 20 protects against Extra 25
East and Extra 25 West between C and D
after two thirty 230 PM

May Work Extra 20 assume that Extra 25 will move eastward before it moves westward?

No. It must be protected against the extra in both directions not later than 2:30 p.m.

ORDER No. 127

Eastward Extra Trains wait at C until
330 PM

Assuming that Work Extra 20 receives Order No. 127 at C and holds orders to work between C and D, can it leave C and move eastward?

No. Work Extra 20 is an "eastward extra train" when moving eastward; and is therefore held at C by the 3:30 p.m. wait.

How should Order No. 127 have been worded?

"Eastward Extra Trains except Work Extra 20 wait at C until 330 PM."

ORDER No. 128

Eng 33 works extra 930 AM until 830 PM
between C and D with right over all trains

May any train enter the working limits of Work Extra 33, even under protection of flag?

No.

When is such an order necessary?

Only in case of emergency, such as a washout, slide, derailment, etc., when the work train must have no interference in doing its work clearing the main track.

ORDER No. 129

Eng 20 works extra on Westward track
645 AM until 530 PM between A and C

Under Order No. 129, is work Extra 20 required to clear regular trains?

Yes.

What is required with regard to extra trains?

Whether standing or moving, it must protect itself against extra trains moving with the current of traffic.

ORDER No. 130

Westward Extra trains wait at C until
201 PM

Given Order No. 130, may Work Extra 20 move against the current of traffic on westward track, clearing regular trains?

Yes.

If extra trains are authorized by rule to move with the current of traffic on two or more tracks without running orders, may the dispatcher issue an order such as No. 130, or give a work extra an order not to protect against extra trains?

No.

If Work Extra 20 were authorized to work extra on both tracks between A and C, how must it move?

Unless otherwise provided, it must move with the current of traffic.

If a work extra is given right over all trains on one or more tracks, may it move on those tracks in either direction, the same as when holding a similar order on single track?

Yes. No train may enter its work limits.

ORDER No. 131

Eng 20 works extra on Eastward track
645 AM until 601 PM between D and C
Eastward Extra trains wait at C until
1201 PM

Under Order No. 131, may Work Extra 20 move against the current of traffic, clearing regular trains, until 12:01 p.m.?

Yes.

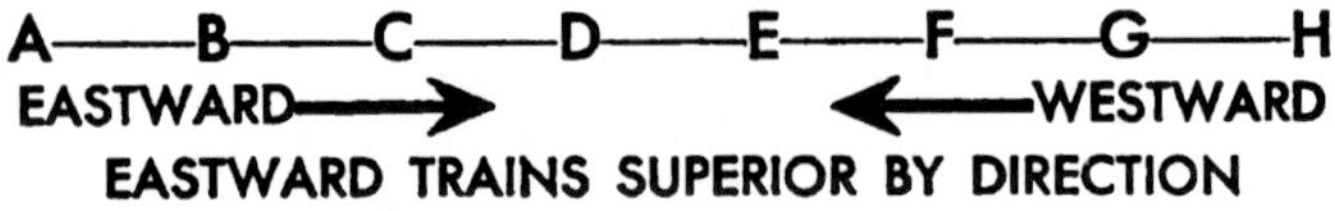

After 12:01 p.m. how must Work Extra 20 be governed?

It must protect itself against extra trains moving with the current of traffic, clearing regular trains.

To whom will annulling orders be addressed?

If an order to be annulled has not been delivered to a train, the annulling order will be addressed to the operator, who will destroy all copies except his own, writing on that, "Annulled by Order No. . . . ." If an order has been delivered to a train, the annulling order will be addressed to the train.

If a conductor has signed for an order but it has not been completed or delivered, must the annulling order be addressed to the train rather than the operator?

Yes.

May the dispatcher add the words, "Order No. 10 is annulled," to an order addressed to a train, omitting these words from the same order addressed to other trains?

No. Rule 202 requires that "Each train order must be given in the same words to all employes or trains addressed."

If a train receives an order reading, "Order No. 20 is annulled," and it has not received Order No. 20, must it be furnished a copy?

Yes.

ORDER No. 132

No. 1 meet No. 2 at E and No. 4 at C

ORDER No. 133

That part of Order No. 132 reading No. 1 meet No. 2 at E is annulled

How should No. 1 be governed with respect to No. 2?

It should clear its time-table time.

After meeting No. 2, how should No. 1 be governed?

It should proceed to C and there meet No. 4.

ORDER No. 134

No. 1 wait at A until 930 AM
B 950 AM
C 1010 AM

ORDER No. 135

That part of Order No. 134 reading C 1010 AM is annulled

Is there anything wrong with Order No. 135?

It does not conform to the Standard Code, although some roads require its use.

How should Order No. 135 be worded to conform to the Standard Code?

"That part of Order No. 134 reading No. 1 wait at C until 1010 AM is annulled."

When a portion of an order is annulled, are other provisions of the order still in effect?

Yes.

ORDER No. 136

No. 1 hold main track meet No. 2 at B

ORDER No. 137

No. 1 meet No. 2 at C instead of B

ORDER No. 138

Order No. 137 is annulled

What is required of No. 1 under Order No. 137?

They must take siding and meet No. 2 at C instead of B.

When Order No. 136 is superseded by Order No. 137, is it void, the same as if annulled?

Yes.

Can Order No. 136 be restored to effectiveness by annulling Order No. 137?

No.

After Order No. 137 is annulled by Order No. 138, where is the meet between No. 1 and No. 2?

There is none.

If No. 2 is due, or overdue, at the point where No. 1 receives Order No. 138, what action must be taken?

They must get out a flag immediately to protect against No. 2.

A——B——C——D——E——F——G——H

EASTWARD——→ ←——WESTWARD

EASTWARD TRAINS SUPERIOR BY DIRECTION

ORDER No. 139

No. 1 meet No. 2 at B

ORDER No. 140

No. 1 meet No. 2 at C instead of B

ORDER No. 141

No. 2 meet No. 1 at B instead of C

Are Orders Nos. 139, 140 and 141 in the prescribed form?

Yes.

Would a dispatcher be using good judgment to issue orders in this combination?

No. Order No. 141 should have read: "Order No. 140 is annulled. No. 2 meet No. 1 at B."

ORDER No. 142

No. 1 meet No. 8 at D instead of C and
hold main track and meet No. 4 at B

How should No. 1 be governed under Order No. 142?

No. 1 should take siding at D and meet No. 8; then hold main track at B and meet No. 4.

ORDER No. 143

No. 2 has right over opposing trains on
Westward track B to D

Before a movement can be authorized against the current of traffic, what is required of the dispatcher?

The track must be cleared of opposing trains.

Does Order No. 143 prohibit No. 2 from using the eastward track between B and D?

Yes.

May opposing trains leave D before No. 2 arrives?

No.

If an extra west is at D and can clear No. 2's schedule time at C, can it go?

No.

If an inferior eastward train is between B and D on eastward track, must it be given a copy of Order No. 143? Why?

Yes. To prevent its having to clear the time of No. 2 between those points.

If an inferior eastward train, which is following No. 2, arrives at D before No. 2 returns to eastward track at that point, must it be given a copy of Order No. 143?

Yes, to prevent it from leaving D ahead of No. 2.

When a section of double track is singled by the use of Form D-S, in which direction are trains superior by direction?

In the direction specified by the time-table.

What is a Block System?

A series of consecutive blocks.

What is a Manual Block System?

A series of consecutive blocks, governed by block signals operated manually, upon information by telegraph, telephone or other means of communication.

What is an Automatic Block System?

A series of consecutive blocks governed by block signals, cab signals, or both, actuated by a train, or engine, or by certain conditions affecting the use of a block.

What is an Interlocking?

An arrangement of signals and signal appliances so interconnected that their movements must succeed each other in proper sequence and for which interlocking rules are in effect. It may be operated manually or automatically.

What are Interlocking Limits?

The tracks between the home signals of an interlocking.

What is a Block Station?

A place from which block signals are operated.

What is an Interlocking Station?

A place from which an interlocking is operated.

What is a Block?

A length of track of defined limits, the use of which by trains is governed by block signals, cab signals, or both.

What is a Fixed Signal?

A signal of fixed location indicating a condition affecting the movement of a train or engine.

What is a Cab Signal?

A signal located in engineman's compartment or cab, indicating a condition affecting the movement of a train or engine and used in conjunction with interlocking signals and in conjunction with or in lieu of block signals.

What is a Block Signal?

A fixed signal at the entrance of a block to govern trains entering and using that block.

What are Interlocking Signals?

The fixed signals of an interlocking.

What is a Home Signal?

A fixed signal at the entrance of a route or block to govern trains or engines entering and using that route or block.

What is an Approach Signal?

A fixed signal used in connection with one or more signals to govern the approach thereto.

What is a Dwarf Signal?

A low home signal.

What is Limited Speed?

A speed not exceeding .... miles per hour.

Note: Where blank spaces are found in the answers to these questions, the student is referred to his own rule book. The Standard Code permits each road to fill in the blank spaces in the rules as they deem best; therefore, the answers to questions will be found in the individual rule book.

If no definition of "Limited Speed" is found in your rule book, then there is no answer to this question as far as your road is concerned. Should "Medium Speed" be defined as a speed not exceeding 30 miles per hour, that would be the answer to the next question. It is recommended that the student determine this point and fill in the blank spaces with correct speeds; or such other information as his rule book prescribes. Most reports of irregularities are made to the train dispatcher, even though the rules may prescribe that they be reported to the superintendent, or other official.

What is Medium Speed?

A speed not exceeding .... miles per hour.

What is Slow Speed?

A speed not exceeding .... miles per hour.

What is Restricted Speed?

Proceed prepared to stop short of train, obstruction, or switch not properly lined, and to look out for broken rail.

May trains run with reference to other trains in same direction by block signals whose indications will supersede the superiority of trains?

Yes, on portions of the railroad, and on designated tracks so specified on the time-table.

Who must be advised in advance, of any known condition that will delay the train or prevent it from making usual speed?

The . . . .

Except as affected by Rule 251 do all Block Signal Rules and Operating Rules remain in force?

Yes.

May trains be governed by block signals whose indications will supersede the superiority of trains for both opposing and following movements on the same track?

Yes, on portions of the railroad, and on designated tracks so specified on the time-table.

Who must be advised in advance of any known condition that will delay the train or prevent it from making usual speed?

The . . . .

Except as affected by Rule 261 do all Block Signal Rules and Operating Rules remain in force?

Yes.

Aspects may be shown by the position of semaphore arms, color of lights, position of lights, flashing of lights, or a combination of color, position, and flashing of lights.

Day and night aspects for color light signals shall have the same colors as the night aspects of the semaphore signals.

Day and night aspects for position light signals shall have the same positions as the day aspects of the semaphore signals.

Aspects shown are typical. Each road should show the aspects and colors of lights it uses.

Note.—In the following illustrations of typical signal aspects, Rules 281 to 292, inclusive:

R = Red
Y = Yellow
G = Green

Note.—When flashing color lights are used, they shall be indicated as follows:

FR = Flashing Red
FY = Flashing Yellow
FG = Flashing Green

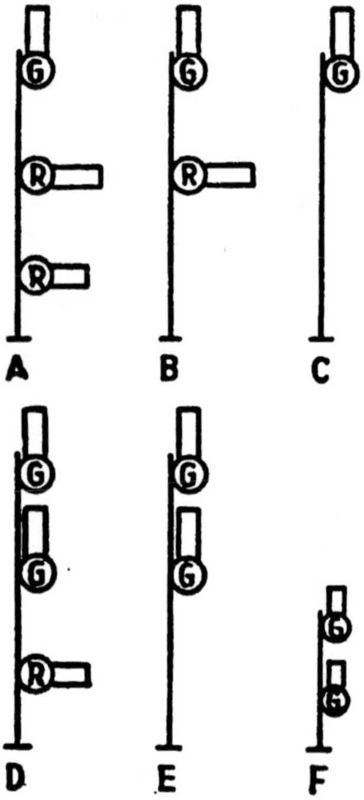

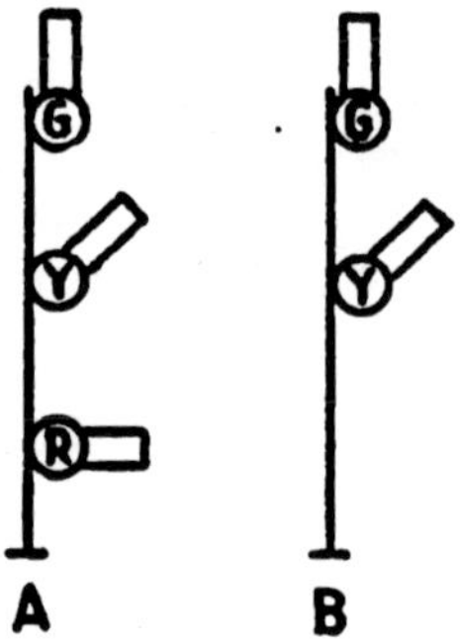

What is this signal indication?

Proceed.

Clear signal.

What is its name?

What is this signal indication?

Proceed, approaching second signal at medium speed.

What is its name?

Advance approach medium.

What is this signal indication?
Proceed, approaching next signal at limited speed.
What is its name?
Approach limited.

What is this signal indication?
Proceed; limited speed within interlocking limits.
What is its name?
Limited—clear.

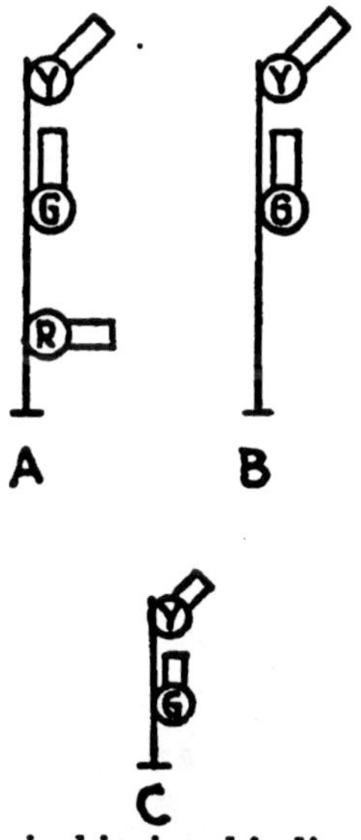

What is this signal indication?
Proceed, approaching next signal at medium speed.
What is its name?
Approach medium.

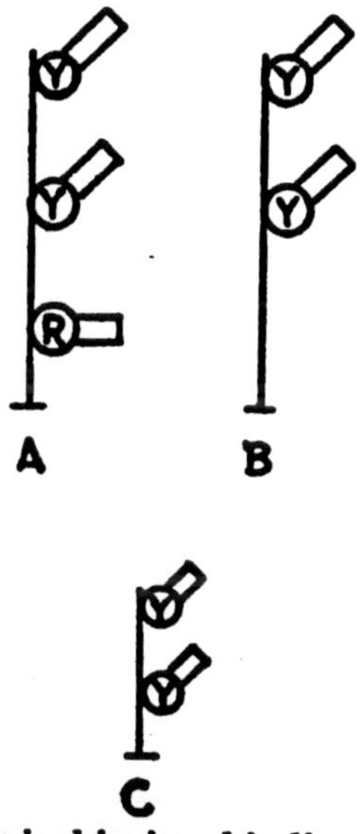

What is this signal indication?
Proceed, preparing to stop at second signal.
What is its name?
Advance approach.

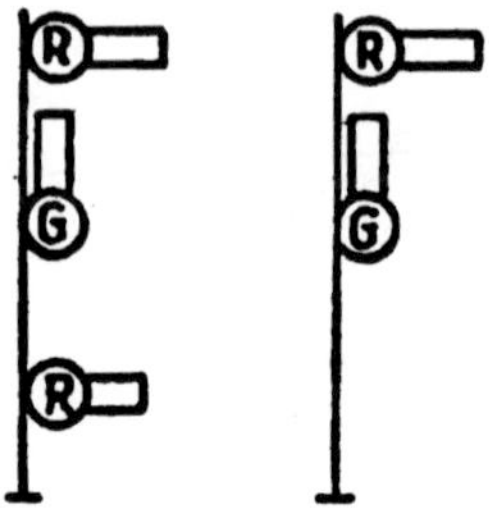

A B

What is this signal indication?
Proceed; medium speed within interlocking limits.

What is its name?
Medium—clear.

What is this signal indication?
Proceed, preparing to stop at second signal; medium speed within interlocking limits.

What is its name?
Medium—advance approach.

What is this signal indication?
Proceed at medium speed, approaching next signal at slow speed.

What is its name?
Medium—approach slow.

What is this signal indication?
Proceed, approaching next signal at slow speed. Train exceeding medium speed must at once reduce to that speed.

What is its name?
Approach slow.

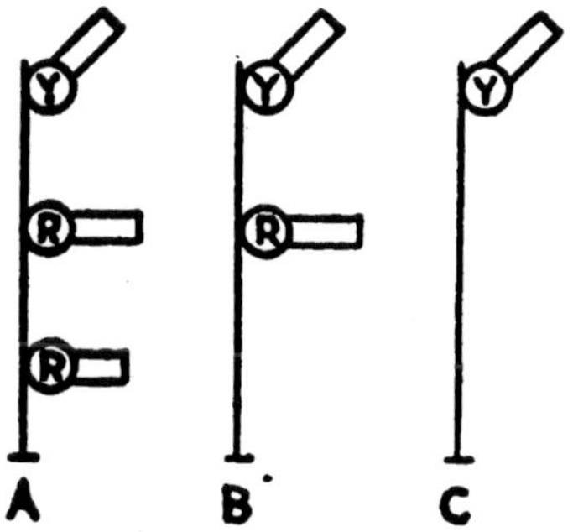

What is this signal indication?
Proceed, preparing to stop at next signal. Train exceeding medium speed must at once reduce to that speed.

What is its name?
Approach.

What is this signal indication?
Proceed at medium speed preparing to stop at next signal.

What is its name?
Medium approach.

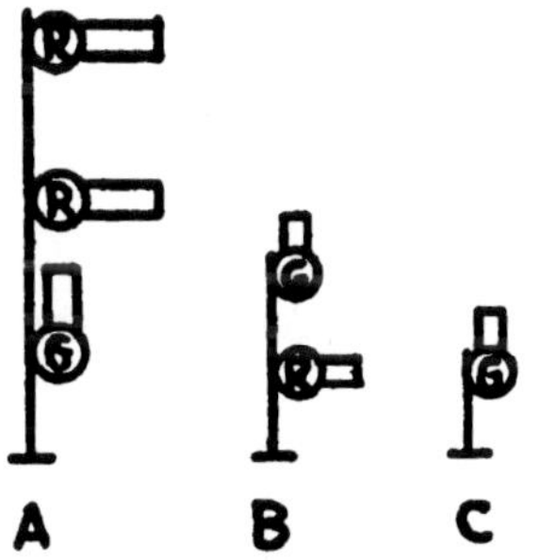

What is this signal indication?
Proceed; slow speed within interlocking limits.

What is its name?
Slow—clear.

What is this signal indication?
Proceed, preparing to stop at next signal; slow speed within interlocking limits.

What is its name?
Slow—approach.

Designate by:

1—Letter plate, or
2—Marker light, or
3—Shape of arm, or
4—Combination of these distinguishing features.

What is this signal indication?
Block occupied; proceed, prepared to stop short of train ahead.

What is its name?
Permissive.

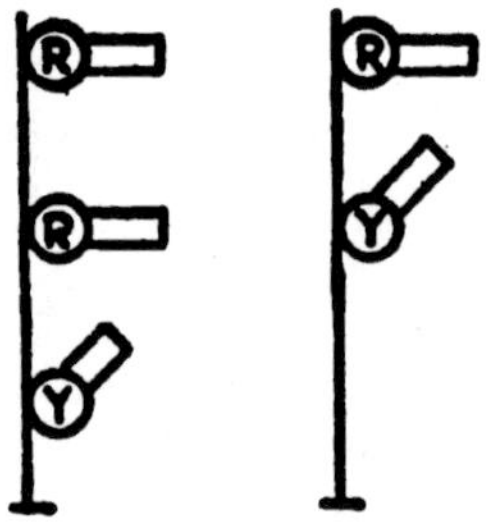

A B

C D

What is this signal indication?
Proceed at restricted speed.

What is its name?
Restricting.

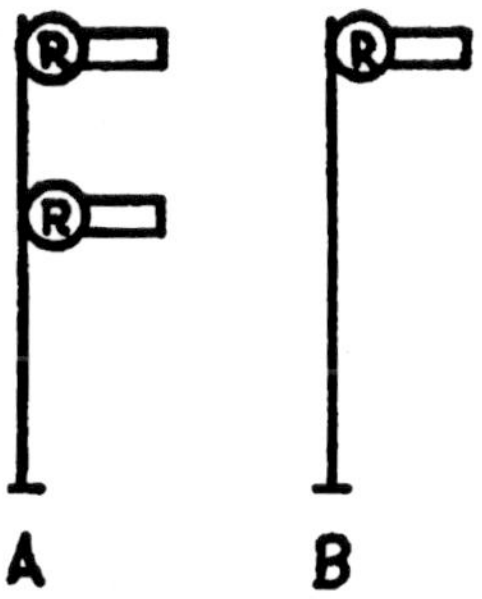

Designate by:

1—Number plate, or
2—Marker light, or
3—Pointed blade, or
4—Combination of these distinguishing features.

What is this signal indication?
Stop; then proceed at restricted speed.

What is its name?
Stop and proceed.

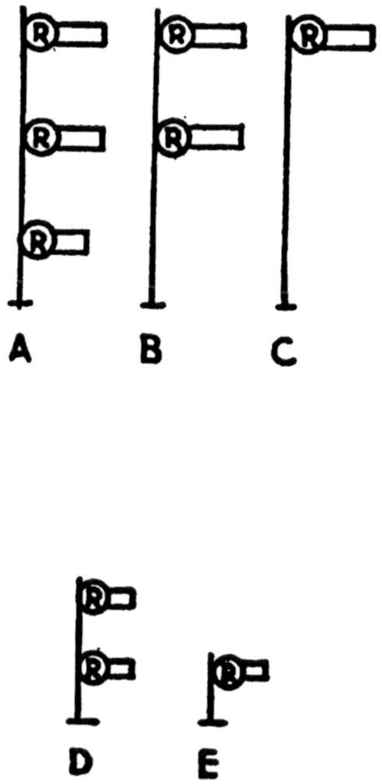

What is this signal indication?
Stop.

What is its name?
Stop.

How must a signal imperfectly displayed, or the absence of a signal at a place where a signal is usually shown, be regarded?

As the most restrictive indication that can be given by that signal.

What is the exception to this requirement?

When the day indication is plainly seen, or when sufficient lights in a position or color position light signal are displayed to determine indication of the signal, it will govern.

When and to whom must a signal imperfectly displayed or the absence of a signal at a place where a signal is usually shown be reported?

Promptly to the ....

Must all members of engine and train crews, when practicable, communicate to each other by its name the indication of each signal affecting the movement of their train or engine?

Yes.

Do block signals govern the use of the blocks?

Yes.

Do they supersede the superiority of trains or dispense with the use or the observance of other signals whenever and wherever they may be required?

No, unless otherwise provided.

What is required when a block station is open at an irregular hour?

Trains must be notified when possible by train order.

May a train having entered a block on other than a proceed indication, accept a proceed indication at any intermediate block station which was closed when such train entered the block?

Not without receiving Clearance Form A.

When Clearance Form A is used will information be shown as to condition of block "clear" or "occupied"?

Yes.

If block is occupied how will train proceed?

Prepared to stop short of train ahead.

Must signals be kept in the position displaying the most restrictive indication?

Yes, except when displayed for an immediate movement.

Must appliances be operated carefully and only by those charged with that duty?

Yes.

What is required when any irregularity affecting their operation is detected?

The signals must be displayed to give their most restrictive indication until repairs are made.

Must signalmen observe, as far as practicable, whether the

indications of the signals correspond with the positions of the levers?

Yes.

May signalmen make or permit any unauthorized repairs, alterations or additions to the apparatus?

No.

When and to whom must any defects in the appliances be reported?

Promptly to the . . . .

Must a block record be kept at each block station?

Yes.

What do the following communicating code numerals indicate:

1?

Keep block signal in stop position for opposing train.

13?

I understand.

2?

Block clear.

5?

Block not clear of train other than passenger.

56?

Block not clear of passenger train.

8?

Opening block station. Answer by 2, 5 or 56.

9?

Closing block station, followed by 2. If the block is clear, answer by 13, followed by 2. If the block is not clear, answer by 5 or 56.

When two or more tracks are used in the same direction, must signalmen, in using the communication code, also specify the track?

Yes.

On single track what is required of the signalman to admit a train to a block?

He must examine the block record, and, if the block is clear of opposing trains and preceding passenger trains, give "1 for . . . .," to the next block station in advance.

What is required of the signalman in advance if it is proper for the train to be admitted?

Reply "2 for . . . ." or "5 of . . . ."

What is then required of the signalman at the entrance of the block?

He must display the proper signal indication.

May a train be admitted to a block which is occupied by an opposing train or by a passenger train except as provided in Rule 333 or by train order?

No.

May a train be permitted to follow a train other than a passenger train into a block under permissive indication or Clearance Form A?

Yes.

(*See the Standard Code not to Rule* 317.)

On two or more tracks what is required of the signalman to admit a train to a block?

He must examine the block record and display proper signal indication.

May a train be admitted to a block which is occupied by a passenger train, except as provided in Rule 333 or by train order?

No.

How may a train be permitted to follow a train other than a passenger train into a block?

Under permissive indication or Clearance Form A.

(*See the Standard Code note to Rule* 318.)

What is required of the signalman when a train enters a block?

He must give train number and the time to the next block station in advance.

What is required of the signalman when the rear of the train has passed . . . . feet beyond the block signal?

He must give the record of the train to the next block station in the rear.

Must this information be entered on the block records?

Yes.

When may signalmen ask for the block?

Not until they have received a report of the train from the next block station in the rear, unless otherwise provided.

Must signalmen, as far as practicable, observe all passing trains and note whether they are complete and in order, and the markers properly displayed?

Yes.

What is required of the signalman should a train pass a block station with any indication of conditions endangering the train, or a train on another track?

He must immediately notify the signalman at the next block station in advance.

What indications must each display to all trains that may be affected?

Stop-indications.

When may each signalman permit any train to proceed?

Not until it is known that its track is not obstructed.

What is required of a signalman when informed of any obstruction in a block?

He must immediately notify the signalman at the other end of the block.

What indications must each display to all trains that may be affected?

Stop-indications.

When may each signalman permit any train to proceed?

Not until it is known that its track is not obstructed.

What is required of the signalman when a train or engine takes a siding or otherwise clears the main track?

He must know that it is clear of the block before giving 2 or displaying a clear-indication for that block.

What is required of the signalman before permitting a train or engine to re-enter the block?

He must obtain control of the block.

What is required of the signalman before permitting a train or engine to enter a block or foul the main track, or to cross over from one main track to another?

He must examine the block record and if all the blocks affected are clear of approaching trains, must provide or arrange for block protection before permission is given.

When may a train be admitted to the block?

Not until movement is complete and block clear, except as prescribed by Rule 289 or Clearance Form A.

Must all crossover movements be entered on the block records?

Yes.

When coupled trains are separated, as prescribed by Rule 364, how must the signalman regard each portion?

As an independent train.

What is required of the signalman when necessary to stop a train for which other than a stop-indication has been displayed and accepted?

He must give hand signals in addition to displaying the stop-indication.

What is required when, from any cause, a signalman is unable to communicate with the next block station in advance?

To stop all trains approaching in that direction.

Should no cause for detaining a train be known, may it then be permitted to proceed with Clearance Form A, provided .... minutes have elapsed since the passage of the last preceding train?

Yes.

May hand signals be used when the proper indication can be displayed by the block signals?

No, except as prescribed by Rule 329 or 343.

When hand signals are necessary, from where and how must they be given?

From such a place and in such a way that there can be no misunderstanding on the part of the enginemen or trainmen as to the signals, or as to the train or engine for which they are intended.

To what trains do block signals for a track apply?

Only to trains moving with the current of traffic on that track.

What is required for blocking trains moving against the current of traffic?

Signalmen will use ....

Who will be responsible for the care of the block station,

lamps and supplies, and, unless otherwise provided, of the signal apparatus?

Signalmen.

How must lights in block stations be placed?

So that they cannot be seen from approaching trains.

If a stop-indication is disregarded, to whom must the fact be reported?

To the next block station in advance and then to . . . .

What is required of the signalman to open a block station?

He must give 8 to the next block station in each direction and record the trains that are in the extended block.

What is required of the signalman when trains, which were in the extended block when the block station was opened and which had passed his block station before it was opened, clear the block in advance?

He must repeat the record to the block station in the rear.

Except upon the authority of . . . . may a block station be closed?

No.

May a block station be closed until the block in each direction is clear of all trains?

No, unless otherwise provided.

What is required of the signalman to close a block station?

He must give 9 followed by 2 to the next block station in each direction and when he receives 13 followed by 2 enter it on his block record, with the time it is received from each block station; the block signals must then be secured in the clear position, all lights extinguished, and the block wires and, when necessary, circuits arranged to work through the closed block station.

When a block station is open at an irregular hour, what signals must signalmen use in addition to block signals, to give the required indications until all trains have passed which have not been notified by train order that the block station is open?

Hand signals.

May signalmen permit unauthorized persons to enter the block station?

No.

To what trains do block signals for a track apply?

Only to trains moving with the current of traffic on that track.

What is required for blocking trains moving against the current of traffic?

.... will be used.

What is required to permit trains to pass a stop-indication?

Must receive Clearance Form A.

May trains proceed on hand signals as against block signals?

No.

Unless otherwise directed, when two or more trains have been coupled and so move past any block station, where must they be separated and who notified?

At a block station, and the signalman notified.

What is required when a train or engine takes a siding or otherwise clears the main track unless the switch involved is operated by the signalman?

Conductor or engineman must so report to the signalman.

May a train or engine enter a block or foul the main track, or cross from one main track to another, without permission of the signalman?

No.

May a train or engine having passed beyond the limits of a block, back into that block without permission of the signalman?

No.

What is required when there is an obstruction between block stations?

Notice must be given to the nearest signalman.

What is required when a train is stopped by a home or block signal?

The conductor or engineman must, when practicable, immediately ascertain the cause.

What is required when any unusual detention occurs at block stations?

Conductors must report to ....

May a block station be considered as closed, except as provided for by time-table or ....?

No.

Do block signals, cab signals, or both, govern the use of the blocks?

Yes.

Do they supersede the superiority of trains or dispense with the use or the observance of other signals whenever and wherever they may be required?

No, unless otherwise provided.

To what trains or engines do block signals for a track apply?

Only to trains or engines moving with the current of traffic on that track unless otherwise provided.

What is required when a train is stopped by a stop-indication?

It must stay until authorized to proceed and then proceed at restricted speed.

What is required in case of failure of means of communication?

Train may proceed when preceded by a flagman to the next signal displaying a proceed indication.

What is required when a train is stopped by a block signal which is evidently out of order?

Unless otherwise provided, the fact must be reported to ....

Must both switches of a crossover be open before a train or engine starts to make a crossover movement?

Yes.

What is required before either switch is restored to normal position?

The movement must be completed.

Where switch indicators are used, do the indications displayed relieve enginemen and trainmen from protecting their train as required by the rules.

No.

What is required before a train or engine enters on or fouls a main track, or crosses from one main track to another, at bolt-locked switches?

After operating the bolt-lock, the trainman must wait ..... minutes before operating the switch.

What is required at non-bolt-locked switches?

Trainman will operate the switch and wait ...... minutes at the switch before making engine or train movement.

Will this relieve employes from the duty of promptly and properly protecting the movement?

No.

What is required of a train or engine that has accepted a clear indication and is delayed in the block?

It must proceed at restricted speed to the next signal.

What is required of a train or engine entering a block between signals?

It must be protected as required by the rules and must proceed at restricted speed.

May a train or engine, having passed beyond the limits of a block, back into that block, except under protection as prescribed by Rule 99 or ....?

No.

Should cab signal and fixed signal indications conflict, which indication will govern?

The more restrictive.

Do cab signal indications supersede fixed signal indications?

No, except when cab signal changes to a more restrictive or a more favorable indication after passing a fixed signal.

What is required, if after passing a fixed signal, the cab signal indication changes from restricted speed to a more favorable indication?

Speed must not be increased until train has run its length.

What do interlocking signals govern?

The use of the routes of an interlocking.

Do their indications supersede the superiority of trains for movements within home signal limits?

Yes.

Do their indications dispense with the use or the observance of other signals whenever and wherever they may be required?

No.

What is meant by one long blast of emergency signal?

All trains and engines within interlocking limits stop immediately.

Two short?

Resume normal movement after receiving the proper signal or permission from the signalman.

Three short?

Whistle or horn test.

Four short?

Call signal maintainer or repairman.

Must signals be kept in the position displaying the most restrictive indication, except when displayed for an immediate movement?

Yes.

Must appliances be operated carefully and only by those charged with that duty?

Yes.

What is required when any irregularity affecting their operation is detected?

The signals must be displayed to give their most restrictive indication until repairs are made.

When the route is set, when must the signals be operated?

Sufficiently in advance of approaching trains to avoid delay.

When necessary to change any route for which the signals have been cleared for an approaching train or engine, when may the switches, movable frogs or derails be changed or signals cleared for any conflicting route?

Not until the train or engine for which the signals were first cleared has stopped.

What is required when any portion of a train or engine is standing on or closely approaching the switch, derail or movable frog?

The levers operating them, also the detector bar or lock, must not be moved.

What is required when a track, switch or signal is undergoing repairs or when a track is obstructed?

Operating levers must be blocked or marked and should not be used.

What is required when the force whose duty it is to keep the switches clear is not on hand when required?

The fact must be reported to . . . .

What is required when a signal fails to work properly?

Its operation must be discontinued and until repaired the signal secured so as to display its most restrictive indication.

Must signalmen observe, as far as practicable, whether the indications of the signals correspond with the position of the levers?

Yes.

May signalmen make or permit any unauthorized repairs, alterations or additions to the interlocking?

No.

Must any defects in the interlocking be reported?

Yes, promptly to the ....

What is required when there is a derailment, or a switch, movable-point frog or derail is run through, or any damage occurs to the track or interlocking?

The signals must be restored so as to display their most restrictive indications.

When may any movement be permitted?

Not until all parts of the interlocking and track liable to consequent damage have been examined and are known to be in safe condition.

What is required when necessary to disconnect a switch, movable-point frog, derail, facing-point lock, detector bar or electric locking circuits?

All switches, movable-point frogs and derails affected must be securely spiked or fastened in the required position and the levers blocked or marked in such a manner that they cannot be operated, before any train or engine is permitted to pass over them.

What is required when switches, movable-point frogs, derails or signals are undergoing repairs?

Stop-indication must be displayed for any movement which may be affected by such repairs, until it has been ascertained from the repairman that the switches, movable-point frogs and derails are properly lined for such movement.

Must signalmen, as far as practicable, observe all passing trains and note whether they are complete and in order?

Yes.

What is required of signalmen should there be any indication of conditions endangering the train, or any other train?

Must take such measures for the protection of trains as may be practicable.

May hand signals be used when the proper indication can be displayed by the interlocking signals?

No.

What is required when the proper indication cannot be displayed?

Hand signals may be used.

What is required when necessary to pass any interlocking signal indicating stop?

Hand signals must be used and . . . . notified.

Who are responsible for the care of the interlocking station, lamps and supplies?

Signalmen.

How must lights in interlocking stations be placed?

So that they cannot be seen from approaching trains.

If a train or engine overruns a stop-indication, to whom must the fact be reported?

To . . . .

May signalmen permit unauthorized persons to enter the interlocking station?

No.

When may an interlocking station be closed?

During a stated period upon authority of . . . .

When so closed, must switches and switch levers be secured for routes that do not conflict?

Yes.

Must signal levers be placed in position so that signals will display the proper indication?

Yes.

Must the interlocking station be securely locked?

Yes.

What is required at interlocking stations, where there is a train order or block signal operated by the same signalman?

Such signals must not be changed to display a proceed indication for a train until after the interlocking signal has been changed to permit the train to proceed.

What is required when a signal indication, permitting a train or engine to proceed, after being accepted, is changed to a stop-indication before it is reached?

Stop must be made at once and such occurrence must be reported to . . . .

What is required to permit trains or engines to pass an interlocking signal indicating stop?

They must receive hand signals.

When may trains or engines proceed on hand signals?

Not until a member of the train or engine crew is fully informed of the situation; the movement must then be made at restricted speed.

May sand be used or water allowed to run over movable parts of an interlocking or spring switch?

No.

What is required when trains or engines are stopped by the signalman in making a movement through an interlocking?

They must not move in either direction until they have received the proper signal from him.

May a reverse movement be made within the limits of an interlocking or a forward movement after making a reverse movement?

Yes, when the proper interlocking signal indication is given or permission is received from the signalman.

How must movements through the interlocking be made while an interlocking station is closed and a signal for an open route indicates "stop"?

Movement must be preceded by a flagman.

What is required of the engineman and trainmen before proceeding?

They must know the route is properly lined, and at the first available point of communication report the facts to the . . . .

Part IV

CENTRALIZED TRAFFIC CONTROL

Centralized traffic control is a trade name for a system devised for dispatching trains without the use of train orders, excepting track orders. Various roads are adopting whatever form of CTC they deem best on their individual properties; and the rules are written on each road to cover the use of the equipment installed. To date there are no A. of A. R. Standard Code CTC rules, although the principles of signal control have been applied to CTC installations.

It will be necessary for the student to study whatever CTC rules have been adopted on the property on which he is employed, as they vary considerably from road to road.

In centralized traffic control territory, the siding switches and signals are remotely controlled, either from a central control board by the dispatcher; or from stations at regular intervals by signalmen, under the direction of the train dispatcher. Two or more tracks are usually signalled in both directions and there is no established current of traffic. Trains may be run on either track in either direction.

All switches not in power control, such as at spurs, leading to the main track are electrically locked; and must be unlocked by the dispatcher before they can be operated.

The signal indications supersede the time-table superiority of trains; that is, a freight train continues to run against, or ahead of a passenger train just so long as the signals indicate proceed. It is the duty of the dispatcher to make all meeting and passing points between both freight and passenger trains.

Instead of trying to remember the provisions of a hatful of orders, the engineman must only obey the indications of the signals. Thus, he gets his instructions just prior to and near the point of execution. In other words, he gets his orders by signal indication at the time and place where they are to be executed.

CTC may be varied, so that certain sidings used infrequently are not equipped with power switches, the switches being operated by the crew when so instructed. The dispatcher illuminates a letter "S" and the rules require that the train operate the switch and enter the indicated track. If more than one track, there will be an "S" for each. In leaving such a track, the dispatcher illuminates the letter "S" at the opposite end, which authorizes the train to open the switch. After the switch has been opened, if the block is clear of trains, the signal will display the proper indication for further movement.

The circuits are so arranged that, should the dispatcher err and attempt to authorize conflicting movements, control of the machine is taken away from him. He cannot authorize conflicting movements by signal indication. Should a train disregard signal indication and overrun a signal, stop signals are automatically displayed in advance for opposing movements to prevent a collision.

One form of CTC is known as A. A. B. S., or Absolute Automatic Block Signals. The circuits here are arranged so as to provide an absolute block for both opposing and following movements.

Another form, known as A. B. S., Automatic Block Signals, does not maintain an absolute block for following movements and includes intermediate signals between absolute signals, so that more than one train may be between stations at one time. As soon as a train passes an intermediate signal, a following train may be admitted to the block on a yellow signal, requiring that they be prepared to stop at the intermediate signal in advance.

Where a considerable number of yard movements must be made on the main track in CTC territory, some of the efficiency of the system is lost, since the train dispatcher finds himself also a yard engine dispatcher.

Switching moves, either by a train or yard engine, can be made in two ways, either by signal indication, the dispatcher handling the switches; or by the dispatcher granting the crew "block authority"; that is, giving them a specified amount of time to work in a given block. He must apply red tags to his levers so that no conflicting movement will be attempted.

Another method of accomplishing the same thing, when several moves are to be made over a given switch, is to allow the crew to take the switch in "hand throw." Moving the mechanism from "power" to "hand throw" automatically sets the signals on either side of the switch to stop to protect the switching movement. When the work is completed, the crew must return the switch to "power" and release it to the dispatcher.

A train or engine which is given time in a block should always leave a portion of its cars on the main track. If a movement enters a spur track, for instance, and takes all its cars, closing the switch, it can make no further movement without authority from the dispatcher.

A graph on the control machine records the time each movement passes certain "OS" sections, such as at each end of a siding. In addition, the dispatcher usually keeps the conventional train sheet.

Leaving a terminal, after a train has its air test and is ready to depart, the dispatcher is notified, and not before. Having received a leaving signal the train is thereafter governed entirely by signal indication. Should some unforeseen delay occur after receiving the signal, the dispatcher must be notified immediately.

There are presently two general forms of CTC. One is similar to a series of consecutive interlockers, where the operator actually handles the switches and signals under the direction of the dispatcher. Multiple tracks may or may not be signaled for traffic in both directions.

The other form of CTC, which seems to be more favored at this time, has a central control board, the dispatcher handling the switches and signals over the entire district by remote control. Some roads have both, and their rules are so worded as to cover either type of installation.

These rules, owing to the newness of this type of dispatching, vary considerably from road to road, and are in a state of revision almost continuously. On some roads, an absolute signal in proceed position only authorizes movement from that signal to the next "A," or absolute signal. The train does not know whether any further movement will be authorized until it actually sees the indication of the next "A" signal, although there is usually an approach signal which indicates the position of the "A" signal.

The general trend is toward applying the A. of A. R. signal rules to CTC to give the signals uniform meaning. In other words, a yellow approach signal requires a train to reduce speed and approach the next "A" signal prepared to stop, instead of only requiring the train to be prepared to stop at the next "A" signal and permitting it to run at full speed up to whatever point the engineer may figure is safe.

It must be realized by the student that a treatise of this kind cannot hope to cover the specific rules written by each road that has made CTC installations. It will be necessary to make a careful study of whatever rules are contained in the individual book of rules, or supplements thereto.

The principle of centralized traffic control is precisely the same as any other method of dispatching trains where signal indication supersedes the time-table superiority of trains. It is reasonable to assume that the same rules will eventually be made to apply to CTC.

There are some roads where the rules provide that signal indication supersedes the superiority of trains. Since a train may be superior to another by right, class or direction, and since right is conferred by train order, theoretically it is impossible for the train dispatcher to make an order effective in such territory.

Some roads do and some do not provide in their rules any method of suspending CTC control and moving trains by train order. In actual practice, it is usually less delay to let trains stand until repairs can be made to CTC equipment than it would be to suspend CTC operation and go to train orders. As a rule, there are few operators left in territory where trains are moved by CTC with a central machine manipulated by the dispatcher.

There is no great uniformity in setting up circuits on some roads. One switch may have a time element of, say, three minutes; or a time lag of that duration. The switch will not unlock for three minutes after the dispatcher has transmitted the impulse that unlocks it. Another switch may require a most complicated set of maneuvers on the part of the dispatcher before it will unlock at all. Still another switch will unlock the moment a trainman takes out the lock and with no action at all on the part of the dispatcher. At another point a switch may not unlock unless cars or an engine are sitting within a certain circuit.

This lack of uniformity arises out of the fact that a CTC installation is usually tailor-made to fit the property on which it is being installed, and according to the ideas of the officers of that property as to the best operating methods.

Therefore, efficient operation is largely dependent on educating operating employes in the correct use of equipment installed on the individual property.

To give the student some idea of the rules under which CTC dispatching is done, the rules of the most recently revised book of rules at hand will be reproduced here. This set of rules, taken from the Uniform Code adopted by several roads including the Missouri Pacific, seems to be about as complete as any yet evolved.

It will be necessary to keep in mind that on these roads ABS refers to conventional automatic block signals, not to any form of CTC. Also, that both types of CTC are in operation on these properties.

CONSOLIDATED CODE OF OPERATING RULES

FORM Z

(1) Effective (Time) (Date) ABS temporarily discontinued from Signal No. .... at (Station) to Signal No. ...... at (Station) Be governed by Rule 356

(2) Effective (Time) (Date) ABS and operation by block signals discontinued from Signal No. .... at (Station) to Signal No. .... at (Station) Be governed by Rule 356

(3) Order No. .... is annulled ABS restored to service

(4) Order No. .... is annulled ABS and operation by block signals restored to service

Example (1) will be used when discontinuing ABS.

Example (2) will be used when discontinuing both ABS and operation by block signals.

Example (3) or (4) will be used when restoration is made.

356. TAKING SIGNALS OUT OF SERVICE.—When emergency requires, ABS and mechanism for movement of trains and engines by block signals, may be taken out of service temporarily by use of train order, Form Z, where territory involved is more than 5 miles, until repairs can be made.

Trains will be operated by time-table, train orders and rules governing such operation. ABS and rules governing movement by block signals will be considered suspended during the time specified and in territory designated.

Block-signal color lights must be extinguished by signal maintainer as soon as possible.

Interlocking rules and interlocking signals are still effective.

Trains must approach railroad crossings, drawbridges, junctions, interlockings and first signal left in service, prepared to stop.

Facing-point spring switches must be examined on ground by throwing over and back by hand.

Remote-control switches must be placed in hand operation. On dual control switches selector lever must be set in "Hand" position, switch operated once by hand and left lined for the main track. At other remote control switches, the switch points must be spiked for main-track movement.

Train order, Form X, must be issued prescribing such speed restriction of passenger trains and freight trains as will insure absolute safety.

Even though ABS rules are suspended, a light burning RED, or a semaphore arm in horizontal position, unless covered, on a block signal must be respected by train stopping, then proceeding at speed prescribed in Form X order.

When the failure of communication in storms, etc., renders it impracticable to deliver train order, Form Z, to trains in ABS territory (not including territory where rules governing opposing and following movements by block signals are in effect), the superintendent, after specifying speed restrictions, may authorize trains having right or schedule that permits them to proceed to consider ABS rules suspended between specified block signals, and in such case all requirements of this rule except issuance of train order, Form Z, will be effective.

Note: (Form X order is for "Speed Restricting or other restrictive conditions")

400. MOVEMENT BY SIGNAL INDICATION.—Within defined limits on designated tracks, so specified on the time-table, or by special instructions, the movement of trains and engines will be governed by block signals whose indications will supersede the superiority of trains for both opposing and following movements on the same track, but do not supersede train orders. Train orders, except Form X orders, are not required within such territory.

Unless authorized by signal indication, trains or engines must not enter a main track at a hand-operated switch without authority of the control operator including track and time limits per Rule 402.

Such operating rules, interlocking rules and automatic block signal rules as are not modified by these rules remain in force.

The movement of trains and engines will be supervised by the train dispatcher, who will issue instructions to the control operator, when required.

401. CLEARANCES AND CLASSIFICATION SIGNALS.—Clearances may be used to designate a train or engine. Classification signals are not required, except when trains are designated by clearance as an extra train or as a section. They may be displayed to avoid stops to place or remove them.

Clearances designating extra trains must read, for example: "Extra 788 South," and clearances designating sections must read, for example: "First 1, Green Signals," "Second 1, No. Signals," etc.

402. TRACK AND TIME LIMITS.—Trains or engines may occupy the main track within specified limits for time periods authorized by control operator specifying track and time limits and track or tracks to be used (to be worded for example: "Track and time limits granted on North Track 1:10 A.M. until 1:25 A.M. between north and south switches of AB siding," or "between Signal No. 625 and Signal No. 655").

While occupying track limits within time granted, trains and engines may move in either direction without flag protection, but must move at low speed, expecting to find trains and engines therein.

A train or engine granted track and time limits, after stopping, may pass a block signal indicating "Stop" or "Stop, Then Proceed at Low Speed" and then proceed at low speed:

(1) to enter track and time limits;

(2) within track limits.

Rule 104 (a), Rule 104 (c), and interlocking rules and interlocking signals must be observed.

Trains and engines must be clear, and reported clear, before expiration of the time granted.

If not clear by the time specified, protection must be afforded in both directions as prescribed by Rule 99.

If additional time is required, authority must be obtained from control operator before authorized time limit has expired.

Control operator must be notified when trains and engines are clear of the track limits granted, except when control operator authorizes by signal indication, a train or engine to move out of the track limits in the same direction in which it entered, it will be considered clear when it has passed such signal indication.

To hold track limits for the time authorized on track or tracks specified, such track or tracks must be occupied continuously, or a main-track switch left open.

No movement may be made under this rule until enginemen have received and understand the track and time limits granted.

When a train or engine requests track and time limits, employe will state his name, occupation, location and train or engine number, and will repeat track limits and time granted, to the control operator, who will then give his OK.

Note: Rules 104 (a) and 104 (c) referred to above will be found on page .

403. REPORTING DELAYS.—Control operator must be advised in advance of any known condition that will delay the train or engine, or prevent it making usual speed.

After a signal indication has been displayed authorizing movement of a train or engine, if movement cannot be made promptly, control operator must be notified immediately.

When a train or engine is stopped by a stop-indication and there is no evidence of an approaching train or engine, communicate promptly with control operator.

404. REVERSE MOVEMENTS.—If it becomes necessary to reverse the movement of a train or engine, such movement must be made under flag protection to the next absolute signal, except during the time and track limits prescribed, or when absolute signal indication is displayed for reverse movement.

If any part of a train or engine passes an absolute signal, authority must be obtained from control operator before reverse movement is made, except as provided by Rule 402.

Should a train or engine overrun an absolute signal displaying stop-indication:

(a) Protect ahead immediately as prescribed by Rule 99.

(b) Communicate with control operator before movement is made in either direction.

405. OPEN THE SWITCH SIGNALS.—When signal governing movement from main track to a siding or nonsignaled track displays indicate "Stop, Open the Switch," per Rule 292-A, train must enter siding or nonsignaled track, after switch is open.

406. CALL LIGHTS.—When a train or engineman of a train or engine which is switching or standing, finds call light burning on relay house or telephone booth, he must communicate immediately with control operator.

425. MOVEMENT REMOTE-CONTROL SIDINGS IN TIME-TABLE AND TRAIN-ORDER TERRITORY.—Movement on main track between switches

of sidings equipped with remote-control switches and signals within time-table and train-order territory, as well as movement into such sidings, will be governed by signal indications which will supersede superiority of trains.

Such locations will be designated by special instructions.

50. MOVEMENT SIGNAL INDICATIONS, ONE DIRECTION.—Within defined limits on designated tracks, so specified on time-table, or by special instructions, the movement of trains and engines with reference to other trains and engines in the same direction, will be governed by block signals, whose indications will supersede the superiority of trains, but do not supersede train orders.

Train orders, except Form X orders, are not required for the movement of trains and engines in specified direction by indication of block signals in this territory.

The movement of trains and engines will be supervised by the train dispatcher.

51. AUTHORITY TO ENTER MAIN TRACK.—Trains or engines must obtain authority of train dispitcher (except in yard limits where special instructions will govern), and comply with requirements of Rules 99 and 104 (15) before entering main track.

A train or engine must not cross over to, or obstruct another main track, without permission of train dispatcher and protection as prescribed by Rule D-99.

52. REPORTING DELAYS.—The train dispatcher must be advised in advance of any known conditions that will delay the train or engine, or prevent

it making usual speed. When work is to be done which may delay the train or engine, permission must be obtained from the train dispatcher before entering the block in which the work is to be done.

Trains or engines instructed by train dispatcher to clear main track for following trains will keep closely advised of trains to be cleared to avoid delay.

When clear of main track, if following train does not pass promptly, communicate with train dispatcher.

453. CLEARANCES AND CLASSIFICATION SIGNALS.—Clearance may be used to designate a train or engine. Classification signals are not required, except when trains are designated by clearance as an extra train or as a section. They may be displayed to avoid stops to place or remove them.

Clearances designating extra trains must read, for example: "Extra 788 South," and clearances designating sections must read, for example: "First 1, Green Signals," "Second 1, No Signals," etc.

Note: The student is reminded that the foregoing rules are not Standard Code rules. The rules of other roads using some form of centralized traffic control may or may not coincide with those presented. Likewise, the questions and answers to follow are based on the rules quoted and are NOT to be taken as authentic answers on any road except those roads that have adopted the rules of the "Uniform Code of Operating Rules."

Within defined limits on designated tracks, so specified on the time-table, or by special instructions, how will the movement of trains and engines be governed?

By block signals.

Will such indications supersede the superiority of trains for both opposing and following movements on the same track?

Yes.

Do such block-signal indications supersede train orders?

No, but train orders, except Form X orders, are not required within such territory.

Unless authorized by signal indication, how may trains and engines enter main track at a hand-operated switch?

Not without authority of the control operator, including track and time limits per Rule 402.

Do such operating rules, interlocking rules and automatic block signal rules as are not modified by these rules, remain in force?

Yes.

By whom will the movement of trains and engines be supervised?

By the train dispatcher, who will issue restrictions to the control operator, when required.

May a clearance be used to designate a train or engine?

Yes.

How must clearances designating sections or extra trains read?

For example: "First 1, Green Signals," "Second 1, No Signals," "Extra 798 South," etc.

Are classification signals required, except when trains are designated by clearance?

No, but they may be displayed to avoid stops to place or remove them.

May trains or engines occupy the main track within specified limits for time periods authorized by control operators specifying track and time limits, and track or tracks to be used?

Yes.

How will such authority be worded?

For example: "Track and time limits granted on North Track 1 10 a m until 1 25 a m between north and south switches of AB siding," or "between Signal No. 625 and Signal No. 655."

What is the purpose of granting track and time limits within this territory?

To permit a train or engine to move in either direction at Low Speed without flag protection.

While occupying track limits within time granted, must

trains and engines move at Low Speed expecting to find trains and engines within the track limits?

Yes.

May trains and engines move within the track limits during the time granted in either direction without flag protection?

Yes.

May a train or engine granted track and time limits, after stopping, pass a "Stop" or "Stop, Then Proceed at Low Speed," indication to enter track limits, or within track limits, at Low Speed, after complying with Rules 104(a) and 104(c) (examination of switches)?

Yes, except that if such signals are interlocking signals within track limits, interlocking signals and interlocking rules must be observed.

Must trains and engines be clear, and reported clear, before expiration of the time granted?

Yes.

If not clear by the time specified, what must be done?

Protection must be afforded in both directions as prescribed by Rule 99.

If additional time is required, from whom must authority be obtained before authorized time limit has expired?

From control operator.

Who must be notified when trains and engines are clear of the track limits granted?

Control operator, except when he authorizes, by signal indication, a train or engine to move out of the track limits in the same direction in which it entered, in which case it will be considered clear when it has passed such signal indication.

To hold track limits for the time authorized on track or tracks specified, what must be done?

Such track or tracks must be occupied continuously, or a main-track switch must be left open.

May any movement be made under this rule until enginemen receive and understand the track and time limits granted?

No.

When a train or engine requests track and time limits, what is required of employe making such request?

He will state his name, occupation, location and train or

engine number, and will repeat track limits and time granted, to the control operator, who will then give his OK.

Who must be advised in advance of any known condition that will delay the train or engine, or prevent it making usual speed?

Control operator.

After a signal indication has been displayed authorizing movement of a train or engine, what must be done if movement cannot be made promptly?

Control operator must be notified immediately.

When a train or engine is stopped by a stop-indication and there is no evidence of an approaching train or engine, what is required?

Communicate promptly with control operator.

If it becomes necessary to reverse the movement of a train or engine, how must the movement be made?

Under flag protection to the next absolute signal, except during the time and within track limits prescribed, or when absolute signal indication is displayed for reverse movement.

If any part of a train or engine passes an absolute signal, what must be done?

Authority must be obtained from control operator before making reverse movement, except as provided by Rule 402.

Should a train or engine overrun an absolute signal displaying stop-indication, what must be done?

(a) Protect ahead immediately as prescribed by Rule 99.

(b) Communicate with control operator before movement is made in either direction.

When signal governing movement from main track to a siding or nonsignaled track displays indication "Stop, Open the Switch," per Rule 292-A what must train do?

Enter siding or nonsignaled track, after switch is open.

When a trainman or engineman of a train or engine which is switching or standing, finds call light burning on relay house or telephone booth, what must he do?

Communicate immediately with control operator.

How will movement on main track between switches of sidings, equipped with remote control switches and signals within

time-table and train order territory, as well as movement into such sidings, be governed?

By signal indications, which will supersede superiority of trains.

How will such locations be designated?

By special instructions.

Within defined limits on designated tracks, so specified on time-table, or by special instructions, how will the movement of trains and engines, with reference to other trains and engines in the same direction, be governed?

By block signals whose indications will supersede the superiority of trains, but do not supersede train orders.

Are train orders, except Form X orders, required for the movement of trains and engines in specified direction by indication of block signals in this territory?

No.

Who will supervise the movement of trains and engines in this territory?

The train dispatcher.

How may trains or engines enter the main track in this territory?

On authority of train dispatcher before entering the main track (except in yard limits where special instructions will govern).

Must such trains or engines comply with requirements of Rules 99 and 104(15) before entering main track?

Yes.

What is required before a train or engine crosses over to, or obstructs another main track?

Permission of train dispatcher and protection as prescribed by Rule D-99.

Who must be advised in advance of any known condition that will delay the train or engine, or prevent it making usual speed?

Train dispatcher.

When work is to be done, which may delay the train or engine, what is required?

Permission must be obtained from the train dispatcher before entering the block in which the work is to be done.

What will trains or engines do when instructed by train dispatcher to clear main track for following trains?

Keep closely advised of trains to be cleared to avoid delay.

When clear of main track, if following train does not pass promptly, what must be done?

Communicate with train dispatcher.

May clearances be used to designate a train or engine?

Yes.

How must clearances designating sections or extra trains read?

For example: "First 1, Green Signals," "Second 1, No Signals," etc., "Extra 798 South" etc.

Are classification signals required except when a train is designated by clearance as an extra train or as a section?

No, but they may be displayed to avoid stops to place or remove them.

INDEX

A

B

C

D

E

F

Spotted something that isn’t quite right?

Let us know and we’ll fix it for future reprints.
And as a thank you for your diligence, we’ll
send you a free Sothis book of your choice.

publisher@sothis.press

This page intentionally left blank

Made in United States
North Haven, CT
04 November 2026

21775164R00240